POWER ESTIMATION AND OPTIMIZATION METHODOLOGIES FOR VLIW-BASED EMBEDDED SYSTEMS

Power Estimation and Optimization Methodologies for VLIW-based Embedded Systems

by

Vittorio Zaccaria
Politecnico di Milano, Italy

Mariagiovanna Sami
Politecnico di Milano, Italy

Donatella Sciuto
Politecnico di Milano, Italy

and

Cristina Silvano
Politecnico di Milano, Italy

SPRINGER-SCIENCE+BUSINESS MEDIA, B.V.

A C.I.P. Catalogue record for this book is available from the Library of Congress.

ISBN 978-1-4419-5339-1 ISBN 978-0-306-48730-9 (eBook)
DOI 10.1007/978-0-306-48730-9

Printed on acid-free paper

Originally published by Kluwer Academic Publishers in 2003
Softcover reprint of the hardcover 1st edition 2003

To my parents

Foreword

LOW power design is playing an important role in today ultra-large scale integration (ULSI) design, particularly as we continue to double the number of transistors on a die every two years and increase the frequency of operation at fairly the same rate. Certainly, an important aspect of low power faces with mobile communications and it has a huge impact on our lives, as we are at the start-line of the proliferation of mobile PDA's (Personal Digital Assistants), Wireless LAN and portable multi-media computing. All of these devices are shaping the way we will be interacting with our family, peers and workplace, thus requiring also a new and innovative low power design paradigm.

Furthermore, low power design techniques are becoming paramount in high performance desktop, base-station and server applications, such as high-speed microprocessors, where excess in power dissipation can lead to a number of cooling, reliability and signal integrity concerns severely burdening the total industrial cost. Hence, low power design can be easily anticipated to further come into prominence as we move to next generation System-on-Chip and Network-on-Chip designs.

This book is entirely devoted to disseminate the results of a long term research program between Politecnico di Milano (Italy) and **STMicroelectronics**, in the field of architectural exploration and optimization techniques to designing low power embedded systems.

Some of the recent and most promising advances in techniques and methodologies for power-conscious design are presented here, spanning different levels of design abstraction, from SW transformations and algorithms to high performance VLIW processor's micro-architecture, power macro-modeling and optimization.

I'm glad to see that the outstanding work carried out by Dr. Zaccaria during his PhD fits tightly with **STMicroelectronics**' continuous committment on research and innovation.

Andrea Cuomo
Corporate Vice President
General Manager, Advanced System Technology
STMicroelectronics

Preface

This book is a revised edition of my doctoral thesis, sumbmitted in 2002 to the Politecnico di Milano. Although this book represents an overview of my recent research activity on low power design, it can also be viewed as an introduction to the field for those interested in applying power estimation and optimization to their designs.

The main contribution of this book consists of the introduction of innovative power estimation and optimization methodologies to support the design of low power embedded systems based on high-performance Very Long Instruction Word (VLIW) microprocessors. A VLIW processor is a (generally) pipelined processor that can execute, in each clock cycle, a set of explicitly parallel *operations*; this set of operations is statically scheduled to form a *Very Long Instruction Word*.

The proposed estimation techniques are integrated into a set of tools operating at the instruction level and they are characterized by efficiency and accuracy. The aim is the construction of an overall power estimation framework, from a system-level perspective, in which novel power optimization techniques can be validated.

The proposed power optimization techniques are addressed to the micro-architectural as well as the system level. Two main optimization techniques have been proposed: the definition of register file write inhibition schemes that exploit the forwarding paths, and the definition of a design exploration framework that allows an efficient fine-tuning of the configurable modules of an embedded system.

The book is structured as follows. Chapter 2 analyzes the main sources of power consumption of a digital system and the possible abstraction levels at which a system can be represented. Chapter 3 analyzes the state of the art of the literature on power estimation of digital circuits. Chapter 4 describes a general power model for VLIW processors. Chapter 5 applies the proposed power model to a single cluster Lx processor core, while Chapter 6 extends the power analysis to the global system composed of the Lx processor the caches and the main memory. Chapter 7 represents an introduction to the literature on techniques for power reduction while Chapter 8 introduces an hardware optimization technique for pipeline-based microprocessor system. Finally, Chapter 9 shows an integrated framework to perform power exploration at the system level.

Each chapter from 4 to 9 reports and discusses some experimental results that evaluate the effectiveness of the proposed methodologies through the application to industrial case studies. Finally, Chapter 10 reports the concluding remarks and future directions of the research work.

Finally I wish to thank my advisors, Professor Cristina Silvano, Professor Mariagiovanna Sami and Professor Donatella Sciuto for guiding my efforts and for constantly monitoring my progresses during all these years. I am also in debt with them for giving me the opportunity to go around the world to present the results of this research work.

A special thank goes to Roberto Zafalon of STMicroelectronics for giving me the opportunity to apply and validate my ideas on industrial case studies. I am also in debt with him for the huge industrial experience that he transferred to me.

Many special thanks to Professor Luca Benini, Davide Bruni and Guglielmo Olivieri of Universita' di Bologna for their collaboration to the research activity described in Chapter 6.

A special thank to Professor William Fornaciari of Politecnico di Milano for its collaboration to the research activity described in Chapter 9.

My special thanks also to my past and present students of the Politecnico di Milano and of ALaRI, Lugano, for their contribution to many aspects of this research work. Among them, I would like to mention Alessio Lion, Jonathan Molteni, Marco Gavazzi, Andrea Bona, Lorenzo Salvemini and Gianluca Palermo.

VITTORIO ZACCARIA

"But what is it good for?"

Anonymous engineer at IBM's advanced computing systems division commenting on the microchip, 1968

Contents

Preface ix

List of Figures xvii

List of Tables xxiii

1. INTRODUCTION 1
   - 1 Original contributions of this book 2
     - 1.1 Power estimation 2
     - 1.2 Power optimization 4

2. MICROPROCESSOR ABSTRACTION LEVELS 7
   - 1 Transistor abstraction level 9
   - 2 Gate abstraction level 10
   - 3 Register-Transfer abstraction level 11
   - 4 Microarchitecture abstraction level 11
   - 5 Instruction-Set-Architecture abstraction level 14
   - 6 System abstraction level 16

Part I POWER ESTIMATION METHODS

3. BACKGROUND 21
   - 1 Sources of power consumption 21
     - 1.1 Switching power 22
     - 1.2 Short-circuit power 26
     - 1.3 Leakage power 27
   - 2 Power estimation techniques 28
     - 2.1 Transistor-level power estimation 28
     - 2.2 Gate-level power estimation 29

2.3 RT-level power estimation 33
2.4 Microarchitecture-level power estimation 35
2.5 Instruction-level power estimation 36
2.6 System-level power estimation 38

4. INSTRUCTION-LEVEL POWER ESTIMATION FOR VLIW PROCESSOR CORES 41
1 Introduction 42
1.1 Overview of the Work 42
2 System Architecture 46
2.1 Target Processor Architecture 46
2.2 Target ISA 48
2.3 Target Processor Requirements 49
3 VLIW Energy Model 49
3.1 Average Energy per Stage 53
3.2 Average Energy of a Data Cache Miss 54
3.3 Average Energy of an Instruction Cache Miss 54
3.4 Model Reduction for Scalar Pipelined Processors 55
3.5 Model Characterization 55
3.6 Plug-in of the Model into an ISS 57
4 Experimental Results 57
4.1 Design of Experiments 59
4.2 Model Characterization 60
4.3 Model Validation 61
5 Conclusions 62

5. SOFTWARE POWER ESTIMATION OF THE LX CORE: A CASE STUDY 67
1 Introduction 67
2 The VLIW Power Model 68
3 Clustering of the Operations 70
4 Horizontal Scheduling for Low-Power 73
5 The Lx Case Study 75
5.1 Lx Power Characterization 76
5.2 Lx Power Optimization 78
5.3 Early Design Space Exploration 79
6 Conclusions 84

6. SYSTEM-LEVEL POWER ESTIMATION FOR THE LX ARCHITECTURE 87

1	Introduction	88
2	Power Estimation Framework	89
	2.1 Target System Architecture	91
3	Power Macro-Modeling	91
	3.1 VLIW Core Model	91
	3.2 Register File Model	91
	3.3 Cache Model	94
4	Experimental Results	98
5	Conclusions	100

Part II POWER OPTIMIZATION METHODS

7.	BACKGROUND	105
1	Technology/physical-level optimizations	105
2	Logic-level and RT-level optimizations	108
	2.1 Circuit Factorization	108
	2.2 Low power technology mapping	108
	2.3 Gated clocks	109
	2.4 Gate freezing	110
	2.5 Pre-computation	110
	2.6 Finite state machines for low power	111
	2.7 Retiming	112
3	Microarchitecture level optimizations	112
	3.1 Bus Power Reduction	114
4	Instruction-level optimizations	115
5	Design space exploration at the system level	117
6	System level power management	119
8.	A MICRO-ARCHITECTURAL OPTIMIZATION FOR LOW POWER	123
1	Introduction	124
2	Background	125
3	Main contribution of this work	126
4	Low-Power Forwarding Architecture	127
5	Exception Handling and Cache misses	132
6	An Application Example	134
	6.1 Results of Register Liveness Analysis	134
	6.2 Results of Power Analysis	136

7 Extension of the basic idea 138
8 Conclusions 141

9. A DESIGN SPACE EXPLORATION METHODOLOGY 143
1 Design space exploration framework 144
1.1 System description 144
1.2 Design space exploration 145
1.3 Computing the cost of a configuration 146
2 Sensitivity-Based Optimization 148
2.1 Architectural space definition 148
2.2 Sensitivity of the cost function 150
2.3 Sensitivity-based local search algorithm 151
3 The Lx architecture: a case study 155
3.1 Application of the Tuning Phase 157
3.2 Sensitivity-Based Optimization 160
4 The μSPARC architecture: a case study 167
4.1 Application of the Tuning Phase 170
4.2 Sensitivity-Based Optimization 172
5 Conclusions 180

10. CONCLUSIONS AND FUTURE WORK 181
1 Ongoing and future research on power estimation 182
2 Ongoing and future research on power optimization 183

The Mediabench suite 185

References 187

List of Figures

2.1	Possible abstraction levels for the description of microprocessor based embedded systems	9
2.2	Example of a gate-level description of a circuit in Verilog	12
2.3	RT-level description in Verilog of the circuit of Figure 2.2	12
3.1	A CMOS inverter	23
3.2	A rising output transition on the CMOS inverter	23
3.3	A falling output transition on the inverter	24
3.4	A static CMOS NAND gate	26
3.5	Short-circuit current generated during a transition of a CMOS inverter	26
3.6	Logic circuit without reconvergent fanout	30
3.7	Logic circuit with reconvergent fanout	30
3.8	Gate-level power estimation flow	32
3.9	A power state machine for a disk drive	39
4.1	Traditional (a) and pipeline-aware (b) instruction-level energy models	44
4.2	Comparison between traditional and pipeline-aware instruction-level energy models.	45
4.3	Architecture of the target VLIW pipelined core	47
4.4	A trace of execution showing two data cache misses and two instruction cache misses	52
4.5	The system-level test-bench used to validate the model.	58

4.6	Agreement between measured and estimated power values when neither I-cache misses nor D-cache misses occur. (Absolute maximum error 11.4%).	63
4.7	Agreement between measured and estimated power values when D-cache misses occur. (Absolute maximum error 20.2%).	63
4.8	Agreement between measured and estimated power values when I-cache misses occur. (Absolute maximum error 9.6%).	64
4.9	Comparison between measured and estimated power values for the given set of DSP benchmarks. (Absolute maximum error 10%).	64
4.10	Errors generated not considering D-cache miss events	65
4.11	Errors generated not considering I-cache miss events	65
5.1	Similarity in the power cost of some basic Lx operations.	72
5.2	The algorithm for horizontal low power scheduling	74
5.3	The bipartite graph generated from the current and the preceding bundle	75
5.4	Tradeoff between the number of experiments required to characterize the model and the maximum variance within each cluster.	77
5.5	Scatter plot of the measured power values with respect to the estimated power values	78
5.6	Power and energy savings obtained by our horizontal rescheduling algorithm applied to the Lx processor executing the selected set of benchmarks.	79
5.7	Energy-Delay product for the ADPCM encoder benchmark, by varying clusters and cache dimensions	83
5.8	Energy-Delay product for the FIR1 benchmark, by varying clusters and cache dimensions	83
5.9	Energy-Delay scatter plot for the MPEG2 decoder benchmark, by varying clusters and cache dimensions	84
5.10	Energy-Delay scatter plot for the AES encoder and decoder benchmark, by varying clusters and cache dimensions	84
6.1	Power estimation framework.	89

6.2 Agreement between measured transistor-level power values and estimated power values for the Register File (Maximum error within ±8%) 93

6.3 Agreement between measured transistor-level power values and estimated RT-level values for the cache blocks. 96

6.4 Comparison between Instruction-Level power estimates and RTL power estimates for the benchmark set. (Average Error 5.2% - Maximum Error 7.9%). 99

6.5 Break-out of the RTL power contributions due to core, RF, I-cache, and D-cache for the benchmark set. 99

6.6 An example of ISS-based system-level dynamic power profiling capability. 100

7.1 Normalized delay (D) versus the supply voltage (V_{dd} [V]) for a typical gate in a standard CMOS process. 107

7.2 (a) Single clock, flip-flop based finite state machine, (b) clock gated finite state machine 110

7.3 The precomputation architecture 111

8.1 Fragment of code showing a short-lived variable 126

8.2 The proposed low-power 5-stage pipelined VLIW forwarding architecture. 128

8.3 Example of a segment of a 4-way VLIW assembly execution trace of a DCT algorithm 131

8.4 An example of exception handling. 134

8.5 Average cumulative distribution of registers liveness 137

8.6 Cumulative distribution of registers liveness for the selected benchmark set. 138

8.7 Average power consumption of RF read/write accesses (Values are normalized to BaseCost). 139

8.8 Power saving percentage of our optimization approach applied to registers liveness less than or equal to 2 (and 3) clock cycles for the selected set of benchmarks. 140

8.9 Example of a segment of a 4-way VLIW program using checkpointing 141

9.1 The proposed design space exploration flow 145

9.2 The target system architecture. 149

9.3 The algorithm of the tuning phase. 152

9.4 The sensitivity-based local search algorithm. 153

9.5 Normalized delay [*CPI*] vs. energy [*JPI*] scatter plot for the `jpegenc` benchmark. 156

9.6 Delay for *jpegenc* benchmark, by varying cache sizes and block sizes 158

9.7 ED product for *jpegenc* benchmark, by varying cache sizes and block sizes 159

9.8 Delay for *jpegenc* benchmark, by varying cache sizes and associativity 160

9.9 ED product for *jpegenc* benchmark, by varying cache sizes and associativity 161

9.10 Delay for *jpegenc* benchmark, by varying block sizes and associativity 162

9.11 ED product for *jpegenc* benchmark, by varying block sizes and associativity 163

9.12 The distribution of the sensitivity for the set of the architectural parameters ordered by their mean sensitivity 164

9.13 The path followed by the sensitivity-based algorithm when optimizing the `jpeg` encoder application on the **Lx** system 164

9.14 The optimization time speedup obtained by the sensitivity optimizer is, in the average, of two orders of magnitude. 166

9.15 Normalized delay [*CPI*] vs. energy [*JPI*] scatter plot for the `gsm decoder` benchmark. 169

9.16 Normalized delay [*CPI*] vs. energy [*JPI*] scatter plot for the `mesa` benchmark. 170

9.17 Delay for *mesa* benchmark, by varying cache sizes and block sizes 171

9.18 ED product for *mesa* benchmark, by varying cache sizes and block sizes 172

9.19 Delay for *mesa* benchmark, by varying cache sizes and associativity 173

9.20 ED product for *mesa* benchmark, by varying cache sizes and associativity 174

9.21 Delay for *mesa* benchmark, by varying block sizes and associativity 175

9.22 ED product for *mesa* benchmark, by varying block sizes and associativity 176
9.23 The distribution of the sensitivity for the set of the architectural parameters ordered by their mean sensitivity 178
9.24 The path followed by the sensitivity-based algorithm when optimizing the `jpeg` encoder application on the **microSPARC** system 178
9.25 The optimization time speedup obtained by the sensitivity optimizer is always greater than two orders of magnitude. 179

List of Tables

4.1 VLIW operation classes considered for the target VLIW processor 53
4.2 Characterization results for the matrix ν summed over the pipeline stages for all pairs of operations when neither pipeline stalls nor I-cache misses occur. Energy values are given in [pJ]. 60
4.3 Comparison results of measured vs. estimated power values for each pipeline stage and for the interconnection. 60
6.1 Comparison between IL power estimates and RTL power estimates for the benchmark set for the core processor 92
6.2 Comparison between IL and RTL power estimates for the benchmark set for the Register File 94
6.3 Refill behavior of the I-cache 97
6.4 Behavior of the D-cache 97
6.5 Comparison between Instruction-Level power estimates and RTL power estimates for the benchmark set for the caches 98
9.1 The set of benchmarks used for the sensitivity analysis. 165
9.2 The set of optimal parameters and sensitivity indexes for the chosen set of reference applications. 165
9.3 The set of benchmarks used for the validation of the sensitivity approach . 165
9.4 Comparison between the full search optimal configurations (a_{opt}) and the configurations obtained with the sensitivity optimizer ($\hat{a}_{opt}$). 166

9.5 The set of benchmarks used for the sensitivity analysis. 177

9.6 The set of optimal parameters and sensitivity indexes for the chosen set of reference applications. 177

9.7 The set of benchmarks used for the validation of the sensitivity approach . 177

9.8 Comparison between the full search optimal configurations (a_{opt}) and the configurations derived with the sensitivity optimizer ($\hat{a}_{opt}$). 179

Chapter 1

INTRODUCTION

Low-power consumption is one of the crucial factors determining the success of modern, portable and non-portable electronic devices. Mobile computing systems and biomedical implantable devices are just a few examples of electronic devices whose power consumption is a basic constraint to be met, since their operativity in the time domain depends on a limited energy storage.

However, extended battery life is not the only motivation for low power consumption. Due to the increasing integration density and the scaling up of operating frequencies, the overheating produced by power consumption is becoming one of the most important factors determining the duration of the electronic components of the system. Such excessive heating must be reduced by mechanical intervention (cooling) or advanced packaging solutions, affecting another important factor for the success of a product: the overall system cost.

Modern low power electronic devices are also quite often characterized by tight performance constraints. This trend is currently motivated by the demand for mobile, micro-processor based computing devices running medium or high-end software applications. Among these we can find multimedia devices, such as streaming video encoding and decoding appliances, for which the throughput constraint is strictly dependent on the relative power consumption of the device.

In this context, the designer's intervention is essential for taking high-level architectural decisions and providing creative solutions, while the design synthesis, the efficiency and the quality of the overall process is carried out by a set of automatic Computer Aided Design (CAD) tools for power estimation and optimization. In fact, the role of the designer is currently moving from tasks operating at the lowest abstraction levels to

V. Zaccaria et al. (eds.),
Power Estimation and Optimization Methodologies for VLIW - based Embedded Systems, 1-5.

high-level evaluation and optimization actions considering, at the same time, power, performance and overall system cost. This is justified by the fact that operating at the highest abstraction levels makes it feasible the early exploitation of several degrees of freedom by trading-off accuracy and efficiency of estimations.

Modern, increasingly complex electronic systems are strongly influencing this trend and currently much effort is spent, within the CAD community, toward the investigation of high-level modeling and optimization techniques.

1. Original contributions of this book

The main contribution of this book consists of the introduction of innovative power estimation and optimization methodologies to support the design of low power embedded systems based on high-performance Very Long Instruction Word (VLIW) microprocessors. A VLIW processor is a (generally) pipelined processor that can execute, in each clock cycle, a set of explicitly parallel *operations*; this set of operations is statically scheduled to form a *Very Long Instruction Word.*

The proposed estimation techniques are integrated into a set of tools operating at the instruction level and they are characterized by efficiency and accuracy. The aim is the construction of an overall power estimation framework, from a system-level perspective, in which novel power optimization techniques can be validated.

The proposed power optimization techniques are addressed to the micro-architectural as well as the system level. Two main optimization techniques have been proposed: the definition of register file write inhibition schemes that exploit the forwarding paths, and the definition of a design exploration framework that allows an efficient fine-tuning of the configurable modules of an embedded system.

The remaining part of this section describes in detail the power estimation and optimization techniques studied in this book by underlining their research areas as well as the original contributions of this book.

1.1 Power estimation

The first part of this book is devoted to the definition of a power assessment framework focusing on VLIW-based system. This is currently done by modeling the power behavior of the VLIW core and the remaining modules of the system at the instruction level and by plugging the model into an ISS simulator to provide either run-time or off-line power estimates.

Chapter 3 is devoted to the presentation of the fundamental concepts of power consumption in digital systems and the state-of-the-art of the power estimation techniques.

Chapter 4 proposes a new instruction-level energy model for the data-path of VLIW pipelined processors. The analytical model takes into account several software-level parameters (such as instruction ordering, pipeline stall probability and instruction cache miss probability) as well as micro-architectural-level ones (such as pipeline stage power consumption per instruction) providing an efficient instruction-based power estimation with accuracy very close to those given by RT or gate-level simulation approaches. The problem of instruction-level power characterization of a K-issue VLIW processor is $O(N^{2K})$ where N is the number of operations in the ISA and K is the number of parallel instructions composing the very long instruction. One of the peculiar characteristics of the proposed model is the capability to reduce the complexity of the characterization problem to $O(K \times N^2)$. The proposed model has been used to characterize a 4-issue VLIW core with a 6-stage pipeline, and its accuracy and efficiency has been compared with respect to energy estimates derived by gate-level simulation. Experimental results (carried out on embedded DSP benchmarks) have demonstrated an average accuracy of 4.8% of the instruction-level model with respect to the gate-level model.

Chapter 5 shows the application of the previous instruction-level energy characterization methodology to the Lx processor, an industrial 4-issue VLIW core jointly designed by HPLabs and STMicroelectronics. The characterization strategy takes into account several software-level parameters (such as instruction ordering, multiple issue of operations, pipeline stall probability and instruction cache miss probability) and provides an effective and accurate instruction-level power model by means of instruction clustering techniques. The accuracy of the proposed model has been quantified against a industrial gate-level simulation-based power estimation engine by estimating instruction statistics at the RT-level. The obtained results show an average error of 1.9% between the instruction-level model and the gate-level model, with a standard deviation of 5.8%. Furthermore, the model has been used to define a brand new horizontal execution-scheduling algorithm, with an average energy saving of 12%, and to perform an early design space exploration of a multi-cluster version of the Lx.

Chapter 6 describes a framework for modeling and estimating the power consumption at the system-level for embedded architectures based on the Lx processor. The main goal is to define a system-level simulation framework for the dynamic profiling of the power behavior during an ISS-based simulation, providing also a break-out of the power contributions due to the single components of the system (mainly, the core, the register file and the caches). In this chapter we show how the plug-in of the Lx core power model into a real ISS provokes a loss in accuracy when instruction-related statistics are gathered by means of ISS simulation and not by RTL simulation. Experimental results, carried out over a set of benchmarks for embedded multimedia applications, have demonstrated an average accuracy of 5% of the ISS-based engine with respect to the RTL engine, with an average speed-up of up to four orders of magnitude.

1.2 Power optimization

The second part of this book is devoted to the definition of power minimization techniques operating at the micro-architectural and at the system-level. Two main topics have been devised: *(i)* the definition of register file write inhibition scheme that exploits the forwarding paths, and *(ii)* the definition of a design exploration framework to enable an efficient fine-tuning of the configurable modules of an embedded system.

Chapter 7 is devoted to the presentation of the state-of-the-art of the power optimization techniques.

Chapter 8 proposes a new low-power approach to the design of embedded VLIW processor architectures based on the forwarding (or bypassing) hardware which provides operands from inter-stage pipeline registers directly to the inputs of the function units. The power optimization technique exploits the forwarding paths to avoid the power cost of writing/reading short-lived variables to/from the Register File (RF). Such type of optimization is justified by the fact that, in application-specific embedded systems, a significant number of variables are short-lived, that is their liveness (from first definition to last use) spans only few instructions. Values of short-lived variables can thus be accessed directly through the forwarding registers, avoiding write-back to the RF by the producer instruction, and successive read from the RF by the consumer instruction. The decision concerning the enabling of the RF write-back phase is taken at compile time by the static scheduling algorithm. This approach implies a minimal overhead on the complexity of the processor control logic and no critical path increase. As a matter of fact, the application of the proposed solution to a VLIW embedded core, when accessing the Register File, has shown a power saving up to 26.1% with respect to the unoptimized approach on the given set of target benchmarks. The proposed technique can be extended by introducing a new level on the memory hierarchy consisting of the pipeline microregisters level. This extension would virtually increase RF space and, thus, reduce register spilling code.

Chapter 9 describes a new system-level design methodology for the efficient exploration of the architectural parameters of the memory sub-systems, from the energy-delay joint perspective. The aim is to find the best configuration of the memory hierarchy without performing the exhaustive analysis of the parameters space. The target system architecture includes the processor, separated instruction and data caches, the main memory, and the system buses. To achieve a fast convergence toward the near-optimal configuration, the proposed methodology adopts an iterative local-search algorithm based on the sensitivity analysis of the cost function with respect to the tuning parameters of the memory sub-system architecture. The exploration strategy is based on the Energy-Delay Product (EDP) metric taking into consideration both performance and energy constraints. The effectiveness of the proposed methodology has been demonstrated through the design space exploration of a real-world case study: the optimization of the memory hierarchy of a Lx-based system and a MicroSPARC2-based system executing the set of Mediabench benchmarks for multimedia applications. Experimental results have shown an optimization speedup of two orders of magnitude with respect to the full search approach, while the near-optimal system-level configuration is characterized by an average distance from the optimal full search configuration in the range of 16%.

Chapter 2

MICROPROCESSOR ABSTRACTION LEVELS

The design of electronic systems is quite complex since they usually consist of a large number of components that are heterogenous in nature and include interfaces to different technologies (optical, analog, micromechanical etc.). These components can require different design techniques, and the interaction among them has to be considered too.

The rapid evolution of the microelectronic technology toward smaller mask-level geometries, leading to higher integration levels and higher performance, is also motivating new approaches to the design of digital systems. The scenario offered by CAD methodologies is rapidly evolving to deal with the integration and the complexity levels imposed by the system-on-a-chip (SOC) approach, which aims at integrating in a single VLSI circuit one or more microprocessors and the heterogeneous functional modules that compose the system.

In this dissertation, we use the term *microprocessor-based system* to indicate a system that can execute a set of general-purpose functions by means of a microprocessor, an executable program and any other hardware/software module needed to execute the target functions. The comprehensive view of all these components is referred to as the *system-level* view.

System-level design is the activity that leads to a completely functioning microprocessor-based system both in terms of hardware and software functionalities. This activity is usually composed of two separate tasks, i.e., the design of the *hardware* partition and the *software* partition of the system. The coupling of these two activities can be slightly tight, as in the case of traditional general-purpose system design, or very tight, as in the case of platform-based design [1] and ad-hoc SOC embedded design. In the last case, the major parts of the design are carried

V. Zaccaria et al. (eds.),
Power Estimation and Optimization Methodologies for VLIW - based Embedded Systems, 7-18.

out together by using unified languages and design methodologies both for hardware and software partitions, leading to a harware/software co-design [2, 3, 4, 5, 6, 7, 8, 9] of the embedded system.

In this dissertation we focus on the design of the hardware partition of a microprocessor-based system, represented by the microprocessor in the context of a traditional design flow by considering also some aspects of traditional system-level design such as global optimization and design space exploration. For what concerns the software partition of the system, we focus on the assembly language level for the description of executable programs. At this level, the software application is described in terms of the effective instructions executed by the microprocessor. Several optimizations can be done at this level of abstraction to optimize the power consumption of the system.

The design process of a microprocessor-based digital system is composed of three main phases:

1 Conceptualization and modeling.

2 Synthesis and optimization.

3 Validation.

The conceptualization and modeling task consists of capturing the system behavior into a model by means of a useful description of its functions. The modeling phase is quite complicated due to the intrinsic heterogeneity of the components to be described and several modeling styles can be adopted, operating at different abstraction levels.

The synthesis and optimization task mainly consists of refining the model, from a higher abstraction model to a lower, more detailed, one. The goal is to optimize some figures of merit (such delay, area and/or power consumption) associated with the circuit description that express its quality and conformance to the target requirements. The overall quality of a digital design can be measured by the satisfaction of the different design requirements, whose priority mainly depends on the application domain.

Finally, the validation process can be carried out at different abstraction levels: it mainly involves the verification of the completeness and correctness of the models produced by the synthesis with respect to the original model.

The abstraction levels [10] for the description of a microprocessor-based system can be decomposed in three partitions that are joint together to form the system-level view (see Figure 2.1). The *software abstraction levels* refer to the set of possible abstractions for the description of the software functionality, from abstract/mathematical notation to the assembly language. The *ad-hoc hardware abstraction levels* refer to the abstraction levels that can be used for the description of particular modules such as memories, hardware interfaces, I/O blocks and application specific integrated circuits, going from transistor-level up to behavioral descriptions. Finally, the abstraction levels going from transistor-level up to the Instruction-Set-Architecture level characterize the "microprocessor abstraction levels" composing the traditional design flow of an embedded microprocessor, we are targeting in this book. Note that, in the case of microprocessors systems, the "behavioral level" is typically substituted with the "ISA" level.

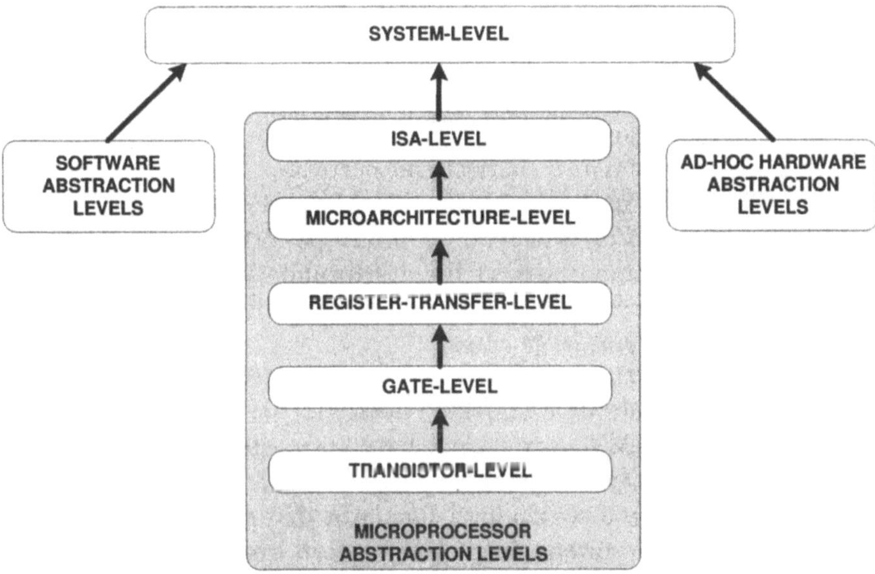

Figure 2.1. Possible abstraction levels for the description of microprocessor based embedded systems

In the following part of the section, we introduce the basic abstraction levels for the description of a microprocessor, from the transistor level up to the ISA and system-level. For each category of abstraction, a brief introduction on the languages and tools used for the description and simulation are be given.

1. Transistor abstraction level

A transistor-level representation describes the operation of the circuit in terms of the electrical behavior of the primitive electrical elements, i.e., transistors, capacitors, inductors and resistors.

A transistor can be modeled at either the device level or the switch level. At the device level, the transistor is described with full non-linear current and voltage analytical expressions, while at the switch level the transistor is approximated as a device with only two modes of operation: "on" and "off". In the "off" state, the transistor is modeled as an open circuit, while in the "on" state, the transistor is modeled by a linear resistance.

SPICE (Simulation Program with Integrated Circuits Emphasis) is probably the most used simulation tool at the device level [11]. SPICE is able to measure effects that could not be captured otherwise with higher level modeling tools such as non-linear capacitances. This is extremely useful for modeling the gate capacitance of transistors, that is actually a function of the operating mode of the transistor and several other parameters [12]. Besides, SPICE allows measurement of voltage and currents involved in the charging and discharging of internal node capacitances and also short-circuit currents.

One of the major drawbacks of such type of simulators is the simulation time. In fact, depending on the performance of the workstation used,and the complexity of the design under test, simulations can run for several hours or days making it unfeasible to perform a transistor level simulation of large chips.

To speed up the circuit simulation, other device level tools use piecewise-linear models stored as tables (CAzM [13]) or table-based event-driven models (PowerMill [14]) achieving 10X to 100X speedup over SPICE with an accuracy always in the range of 10%.

IRSIM [15] is a switch level simulator that relies on completely linear models and two-level voltage quantization to improve simulation speed w.r.t. transistor level simulators. Two-level voltage quantization consists of considering all nodes to be either at V_{dd} or ground rather than some intermediate voltage. IRSIM results to be reasonably accurate for delay estimation of MOS circuits, being about 1000 times faster than SPICE.

In section 2.1, a complete overview of the tools used for transistor level power estimation is given.

2. Gate abstraction level

A gate-level representation describes the operation of the circuit in terms of a structural interconnection of Boolean logic gates (NAND, NOR, NOT, AND, OR and XOR) and flip-flops. Tipically, each basic gate composing the circuit belongs to a *technology library* that provides timing (e.g., delay) and power consumption information about each gate. This representation is usually very accurate with respect to the physical implementation of the circuit, but it strongly depends on the quality of the characterization of the cells done by the technology supplier. Note, however, that, for some blocks of the microprocessor such as the register file and the memory hierarchy, it may be impossible to have an accurate gate level representation since they are usually transitor-level designed (or *custom*) blocks.

A gate-level representation is tipically written by means of either a netlist derived from a schematic or Hardware Description Language (HDL) such as VHDL [16], Verilog [17] and SystemC [18]. These languages support various description levels, including architectural (RT) and behavioural levels. Many tools exist for simulating the descriptions written in VHDL and Verilog such as VSS [19] (for VHDL), VCS [20] and Verilog XL [21] (for Verilog) while, for SystemC, a simulation library is linked with a SystemC description by means of an ordinary C++ compiler to produce a compiled simulation.

3. Register-Transfer abstraction level

The RT-level (or architectural-level) representation usually describes the operation of the circuit in terms of a registers, combinational circuitry, and low low-level buses. This type of representation is useful to express and to analyze the internal architecture of a microprocessor by abstracting from the target library. However, the representation is useful enough to express a cycle accurate behavior of the system

Register-Transfer representations can be described and simulated by means of the same languages and tools such as VHDL, Verilog and SystemC. To give an idea of the compactness of a register-transfer description with respect to a gate-level one, Figure 2.2 shows the Verilog implementation of an example circuit at the gate level while Figure 2.3 shows the same circuit described at the RT-level.

4. Microarchitecture abstraction level

The microarchitecture-level describes how processor interprets the stream of instructions in terms of logic phases of execution and how they impact the logic state of the processor (e.g, the register files and the caches). The phases of execution range from very simple phases, such as the instruction fetch and decode, to more complex ones such as the out-of-order issue phase or the register renaming phase. If different phases, operating on different instructions, are active at the same time, we say that the microarchitecture is *pipelined*. This type of representation is useful to express and to analyze the behavior of the microprocessor

```
...
wire sum3,sum2,sum1,sum0;
wire cout;
xor (sum0, a0, b0);
and (c0, a0, b0);
xor (sum1, a1, b1, c0);
and (t10, a1, b1), (t11, b1, c0), (t12, a1, c0);
or (c1, t10, t11, t12);
xor (sum2, a2, b2, c1);
and (t20, a2, b2), (t21, b2, c1), (t22, a2, c1);
or (c2, t20, t21, t22);
xor (sum3, a3, b3, c2);
and (t30, a3, b3), (t31, b3, c1), (t32, a3, c1);
or (cout, t30, t31, t32);
...
```

Figure 2.2. Example of a gate-level description of a circuit in Verilog

```
...
reg [4:0] sum;
reg cout;
always @(a or b) begin
     sum = a + b;
     if (sum == 0) cout = 1;
     else cout = 0;
end
...
```

Figure 2.3. RT-level description in Verilog of the circuit of Figure 2.2

during the first stages of the design, when an RT-description is not yet available, to obtain rough estimates of the performance of the processor.

Architecture Description Languages [22] are languages designed for description of microarchitecture templates, and are used to perform early validation of systems-on-chip as well as to automatically generate hardware tools (such as simulators) and software tools (such as compilers) required to complete an integrated, and concurrent hardware and software design.

MIMOLA [23] is a synthesis oriented ADL proposed by the University of Dortmund that models the processor structure very similarly to a RT-level language and it is used for both synthesis and compiler generation. However, it does not explicitly support the description of either processor pipelines or resource conflicts.

The UDL-I HDL [24] is a processor description language extensively used within the COACH CAD framework (developed at Kyushu University). The main peculiarities of this language are that it is very close to an RT-level language and that the same processor description can be used for synthesis [25], instruction-level simulation, and compiler construction.

MDES [26] is an ADL for the design space exploration of high-performance processors used within the Trimaran framework [27], a tool for VLIW architectural analysis. It describes both the structure and the behavior of the target processors by means of resource usage, operation latencies, and register file configuration parameters. MDES provides a mean for for describing a hierarchy of memory modules and it captures constraints on ILP by means of reservation tables [28]. MDES has also been used as a description language for VLIW processors belonging to the HPL-PD (Hewelett Packard PlayDoh Architecture) processor family [29].

FlexWare is a CAD system for DSP or ASIP design [30] in which both processor behavior and structure are captured by means of a machine description that consists of three components: instruction set, available resources (and their classification), and an interconnect graph representing the datapath structure. The machine description is then simulated by a VHDL model of a generic parameterizable machine whose parameters are the bit-width, the number of registers and ALUs, etc.

CSDL is a set of machine description languages used in the Zephyr compiler system developed at University of Virginia [31]. CSDL consists of four languages: SLED (for assembly and binary representations of instructions [32]), λ-RTL (for the behavior of instructions in a form of register transfers [33]), CCL (for the specification of function calls [34]) and PLUNGE (a graphical notation for specifying the pipeline structure).

EXPRESSION [35] is an ADL designed to support design space exploration of a wide class of processor architectures (RISCs, DSPs, ASIPs, and VLIWs) coupled with configurable and hierarchical memory system organization. EXPRESSION describes both the structure and the behavior of processor-memory systems by means of a net-list of components (i.e., units, memories, ports, and connections). The pipeline structure is also described by means of ordered units and the timing of the multicycled units. The behavior is described as an instruction-set by means of opcodes, operands, and formats. From an EXPRESSION description, an ILP compiler and a cycle-accurate simulator can be automatically generated.

LISA is a simulation oriented ADL developed by RWTH Aachen that supports bit-true cycle-accurate simulator generation for DSPs [36]. In LISA, both structure and behavior of a processor can be modeled in a pipelined fashion by using complex interlocking and bypassing techniques. Each instruction is described as a set of multiple operations which are defined as register transfers during a single control step. LISA provides explicit support of some pipeline behaviors such as pipeline stalls and flushes by modeling signals which capture the occurrence of certain opcodes in the pipeline stages and trigger different strategies to stall or shift the pipeline. RADL is an extension of the LISA approach proposed by Rockwell Semiconductor Systems [37] supporting a pipeline model characterized by delay slots, interrupts, zero overhead loops, hazards, and multiple pipelines.

Finally, AIDL is an ADL developed by the University of Tsukuba for design of high-performance superscalar processors [38] specially suited for validation of the pipeline behavior, such as for data-forwarding and for out-of-order completion. The behavior of the pipeline is described by means of interval temporal logic and can be simulated by using the AIDL simulator and translated into synthesizable VHDL code.

Microarchitectural level descriptions can be simulated by *architectural* or *performance simulators*. An architectural simulator tracks the microarchitecture state for each cycle, by taking into account instructions "on-the-flight" and simulating faithfully the microarchitecture functions.

Simulation can be either trace-based or execution-driven. A trace-based simulator calculates execution cycles by reading a "trace" of instructions captured during a instruction-set-simulation (see Section 5) and is characterized by a reduced complexity of implementation. An execution-driven simulator "executes" the program, generating a trace on-the-fly by considering the functions and the modules activated by the instructions. Such type of simulator is less-efficient to implement than a trace-based simulator, but it is much more accurate in the case of multiprocessor systems, where trace-based techniques fail to capture interactions among processors.

Among the performance simulators actually in use, SimpleScalar [39], RSIM [40] and MINT [41] are probably the most cited in the literature. Simplescalar provides a set of simulation engines for an out-of-order superscalar processor with a large set of configuration parameters (e.g., memory hierarchy dimensions, number of processor integer units etc...) while RSIM and MINT are execution driven microarchitectural simulators for multiprocessing systems.

5. Instruction-Set-Architecture abstraction level

The instruction-set-architecture-level (ISA-level) describes the processor in terms of the set of registers of the machine and the set of instructions that can be executed on the processor. Tipically, each instruction is described in terms of its effects on the processor state, and it is provided with some rough estimates of the latencies involved in its execution.

The formal description of a processor ISA by means of a description language is useful for the automatic generation of lower level processor descriptions and/or compilers. nML is a language proposed by TU Berlin [42, 43, 44, 45] for the construction of instruction-set retargetable simulators and code generators. nML targets DSPs and ASIPs design by describing the ISA of a processor in the form of an attribute grammar where each instruction is described by a behavior, an assembly syntax, and the object code image for each each instruction. Resource conflicts between instructions are described by sets of legal combinations of operations. ISDL is another ISA description language used for automatic generation of compilers, assemblers, and cycle-accurate simulators [46, 47]. It is used to describe the instruction set of processors and the constraints on ILP posed by resource conflicts.

ISA descriptions can be simulated by means of Instruction-Set Simulators (or ISSs) that can execute a program without a detailed description of the micro-architecture. These simulators can provide access to the internal state of the processor that is hardly visible on the real hardware while executing a program; such information is useful to analyze the behavior of the various system components and to improve the design and implementation of compilers and applications in general.

An ISS can also provide traces of instruction and/or data memory references, register usage patterns, interrupt or exception events and timing statistics (code *profiling*). However, since the ISS does not have a complete timing model of the processor micro-architecture and of the memory hierarchy, the gathered timing statistics are only cycle-approximate.

ISSs can be either *interpreted* or *compiled*. An interpreted simulator, is composed of a loop that reads each instruction of the stream, decodes it and calls a specific function that modifies the simulated processor state and gathers timing statistics. Sometimes, such type of interpreters optimize the code execution by converting the program into an intermediate form (usually *threaded code* [48]) to be interpreted. Among the interpreted simulators we can find SimICS [49] (Motorola 88110 and Sparc v8), Dynascope [50], SPIM [51] (MIPS) and Talisman [52] (Motorola 88K).

A (cross-) compiled simulator translates statically (or dyamically) the instructions of the target architecture in a semantically equivalent code written in the ISA of the host machine. Such type of ISS are much more faster, since there is no fetch and decode loops and the function calls are very limited. However, the emulation of exceptions, interrupts and system calls involves the use of sophisticated techniques and, typically, it does not lead to accurate timing statistics. Example of such type of simulators are Shade [53] (for SPARC by SunMicrosystems) and SimOS [54] (for MIPS).

Among compiled simulators, *augmented* simulators run host instructions native (i.e., the host processor is the same as the target architecture), but some instructions are simulated by means of native code that gathers statistics about particular events (e.g., branches or memory accesses). One of the most important augmented simulators is ATOM [55, 56], a tool for the instrumentation of ALPHA AXP programs based on the object modification library OM [57]. The tool provides a set of functions to generate personalized instrumentation programs to gather arbitrary information about a program execution. Pixie [58] is a program analyzer tool that reads the whole executable for a program and writes a new version of the program that saves profiling information as it runs (allowing the user to collect only basic block execution counts and text and data traces). The Wisconsin Wind Tunnel II (WWT II) [59] is a parallel, discrete-event, multiprocessor simulator based on a augmented ISS that adds instrumentation code to calculate the target architecture execution time and to simulate the target architecture memory system.

Due to the differences cited above, interpreted simulators are well suited for realistic simulations involving complex interactions (e.g., interrupts) between the hardware and the software running on the simulated machine (e.g., a multi-programmed operating system), while a compiled simulator is used typically to perform rough timing optimizations on the application running on the simulated machine in (quasi-)ideal conditions.

6. System abstraction level

A system-level description should be as "expressive" as possible in order to represent both hardware and software functionality at the same time. High-level description languages, such as C or C++, are used to represent both the hardware and software modules and can be used to build executable models useful to derive some estimates of system performance and cost.

A unified system-level description is usually abstracted from the mapping of the hardware and software modules. From such type of specification describing the overall functionality, a mapping to hardware (microprocessor and supporting ASICs) and software modules is performed by partitioning and exploring the architectural design space following a cost/performance figure of merit. The result is another system-level description in which the hardware and software modules are clearly distinguishable and configurable. Among the hardware modules, we can find the microprocessor on which the software is executed and a set of dedicated accelerating co-processors.

A trade-off must be found during the mapping phase since a fully ad-hoc hardware solution, in which all the functionality is performed on dedicated ASICs, may provide higher performances but would rise the cost of the overall system. A fully software solution running only on the microprocessor can lead to a loss of performances but also a reduction in costs.

To respond to the requirements imposed by the system-level design, the worldwide initiative SLDL (System Level Design Language) intends to define an inter-operable language for the high-level specification and design of systems based on micro-electronic components [60]. The mission of the SLDL committee consists of supporting the creation and the standardization of a language in order to describe, to simulate and to verify a digital system based on a single-chip or multi-chip modules.

Recently, SystemC has been proposed as a unified language for system-level modeling. As cited before, SystemC is an hardware oriented language that is capable to capture gate-level, RT-level and behavioral-system level description by means of a set of classes written in C++ language. Clearly, SystemC supports seamlessly software design through standard C/C++ constructs.

Hardware simulation tools are widely available for the simulation of hardware centric systems, whereas only few tools support mixed hardware/software system-level descriptions (e.g, PTOLEMY [61], Mentor Graphics Seamless and CoWare N2C). These tools adopt different simulation strategies for microprocessor based systems. One of these strategies consists of the use of a HDL simulator, with an HDL model for the microprocessor and the ASICs. If a precise real-delay model is adopted, not only the system functionality can be checked, but also accurate timing can be obtained at the expenses of long simulation runs.

A more efficient approach consists of utilizing an instruction-set model of the processor and constructing a cycle accurate wrapper in order to insert the entire model into an hardware (co-)simulation system (see, for example Mentor Graphics Seamless, and CoWare N2C). This approach is much more faster than the previous one and it can enable a reasonably efficient (co-)simulation of the overall system in the early stages of the system development.

The third strategy consists of emulating the hardware components by mapping them into a programmable hardware, as a board of FPGAs. The emulation speed is only one order of magnitude less than the execution on the target hardware, at the expenses of a limited visibility of the internal states. The drawback of this approach is that it can be applied in the late stages of the design, when the ASIC modules and their interfaces with the processor and the software have been completely implemented.

POWER ESTIMATION METHODS

Chapter 3

BACKGROUND

In this chapter we present the main sources of power consumption in digital VLSI circuits based on static CMOS technology. In particular, we briefly analyze the main parameters affecting the total power consumption, namely the clock frequency, the supply voltage, the capacitive load, and the switching activity. Then, we present an overview of the state-of-the-art power estimation techniques that can be used during the design flow of a VLSI-based system.

1. Sources of power consumption

Providing an overview of the sources of power consumption in digital circuits, our focus is on CMOS technology [62, 63].

Although the design of portable devices certainly requires consideration of the peak power consumption for reliability and proper circuit operation, we concentrate on the time averaged dynamic power since it is directly proportional to the battery weight and volume required to operate circuits for a given amount of time.

The average power dissipation of a digital CMOS circuit can be decomposed in a *static* and a *dynamic* component:

$$P = P_{static} + P_{dynamic} \tag{3.1}$$

The static power P_{static} characterizes circuits that have a constant source of current between the power supplies (such as bias circuitry, pseudo-NMOS logic families etc.). Besides, it can become significant in CMOS circuits in the presence of "stuck-at on" transistor faults. However, for properly designed CMOS circuits, this power contribution can be neglected.

V. Zaccaria et al. (eds.),
Power Estimation and Optimization Methodologies for VLIW - based Embedded Systems, 21–40.

The dynamic component $P_{dynamic}$ is the dominant part of the power dissipation in CMOS circuits and, in turn, it is composed of three terms:

$$P_{dynamic} = P_{switching} + P_{short-circuit} + P_{leakage} \tag{3.2}$$

The first term, $P_{switching}$, is called *switching power* and it is due to the charge and discharge of the capacitances associated with each node in the circuit. The second term, $P_{short-circuit}$, is called *short-circuit power* and it derives from the short-circuit current from the supply to the ground voltage during output transitions. The third term, $P_{leakage}$, is called *leakage power* and it is due to the leakage currents which can arise from reverse bias diode currents.

As a matter of fact, the switching power component dominates by contributing up to the 90% of the total power consumption. Thus, the estimation and the minimization of this component is a crucial issue for the design of electronic components and it is the primary target of this work.

However, in rest of this section, we analyze in detail all the three components of the dynamic power.

1.1 Switching power

To simplify the analysis of the sources of the switching power consumption of CMOS circuit, let us consider a simple static CMOS inverter (see Figure 3.1). As in every CMOS gate, the pull-up network is built with PMOS transistors (in our case, T1) and it connects the output node V_{out} to the power supply V_{dd}. Conversely, the pull-down network is composed of NMOS transistors (in our case, T2) and it connects the output node to the ground node V_{ss}. In CMOS gates, the structure of the pull-up and pull-down network is such that when the circuit is stable (i.e. after the output rise or fall transients are exhausted) the output node is never connected to both V_{dd} and V_{ss} at the same time.

The switching power consumption arises whenever the output capacitor C_L of the the CMOS gate is charged through a power supply or it is discharged to the ground. This is due to the finite source-drain resistance of the PMOS and NMOS transistors, providing the path for the charging current.

Let us consider a rising output transition due to a $V_{dd} \rightarrow 0$ input transition (see Figure 3.2) where $i_{dd}(t)$ is the instantaneous current being drawn from the supply voltage V_{dd}. Since when the input is low T_2 is not conducting, the current i_{dd} is absorbed by the capacitor and it is equal to:

$$i_{dd}(t) = C_L \frac{dV_{out}(t)}{dt} \tag{3.3}$$

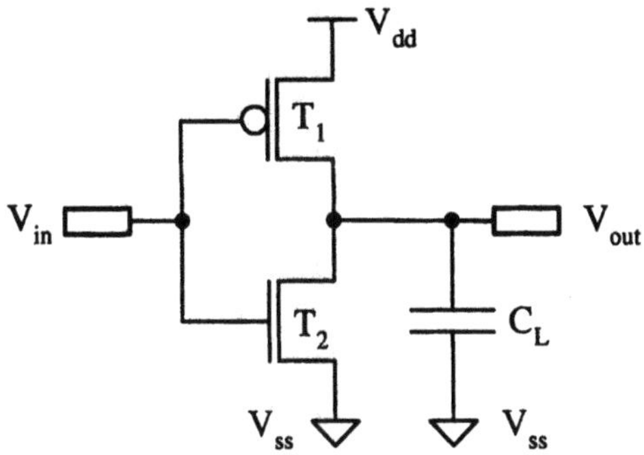

Figure 3.1. A CMOS inverter

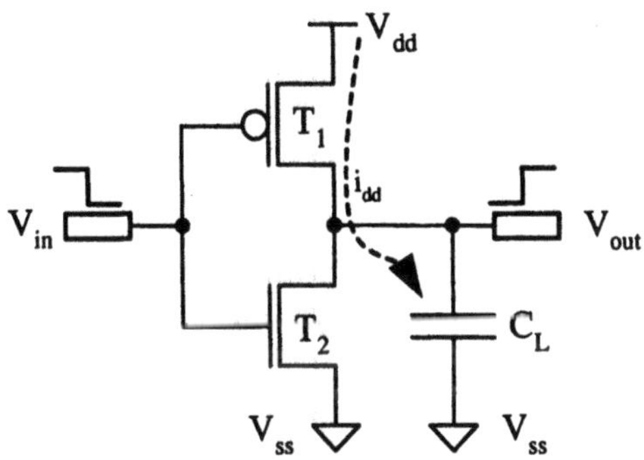

Figure 3.2. A rising output transition on the CMOS inverter

The total energy drawn from the power supply for an output $0 \rightarrow V_{dd}$ transition during a period T can be derived from Eq. 3.3 in the following way:

$$E_s = V_{dd} \int_0^T i_{dd}(t)dt = V_{dd} \int_0^{V_{dd}} C_L dV_{out} = C_L V_{dd}^2 \qquad (3.4)$$

From Eq. 3.4, the energy drawn by the power supply is $C_L V_{dd}^2$ regardless of the waveform at the output of the inverter.

The energy drawn by the inverter can be decomposed in two parts. The first part is dissipated as heat by transistor T_1 while the second part is stored in the output capacitor C_L, being T_2 in the off state.

At the end of the transition, the output capacitor is charged to V_{dd} and the energy stored in it is given by:

$$E_{cap} = \frac{1}{2} C_L V_{dd}^2 \qquad (3.5)$$

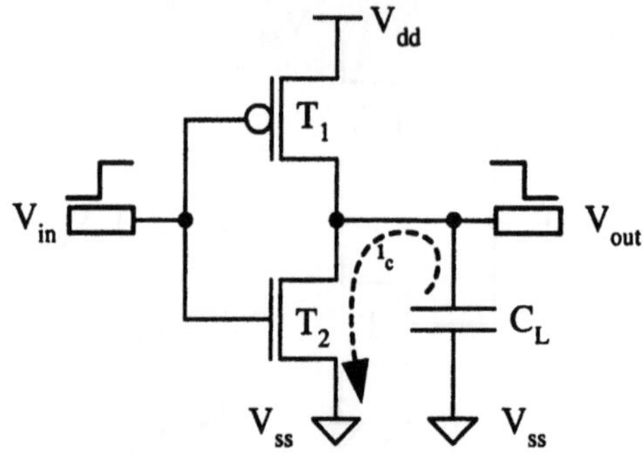

Figure 3.3. **A falling output transition on the inverter**

Therefore, the energy dissipated as heat by transistor T_1, and thus by the gate, during the $0 \rightarrow V_{dd}$ output transition is:

$$E_{0 \rightarrow V_{dd}} = E_s - E_{cap} = \frac{1}{2} C_L V_{dd}^2 \qquad (3.6)$$

If we consider the falling output transition (see Figure 3.3), since no energy is stored in the output capacitor at the end of the transient and T_1 is not conducting, the total energy $E_{V_{dd} \rightarrow 0}$ dissipated by the gate during a falling output transition is given by $\frac{1}{2} C_L V_{dd}^2$.

If we assume one rising/falling transition per clock cycle, then the average power consumption of a gate is given by:

$$P = \frac{E_{0 \rightarrow V_{dd}}}{T} = \frac{E_{V_{dd} \rightarrow 0}}{T} = \frac{1}{2} C_L V_{dd}^2 f \qquad (3.7)$$

where T is the clock period and f is the clock rate ($f = \frac{1}{T}$).

Equation 3.7 expresses the power consumption of the inverter when one output rising or falling transition occur in each clock cycle. The equation can be extended to a more general form that takes into account the probability that a transition could occur in a clock cycle. This probability is called the *output switching activity* (symbolically referenced to as α) and it is the sum of the probabilities that a rising or a falling transition occurs on the output in each clock cycle.

By introducing the switching activity α, Eq. 3.7 can be extended as:

$$P = \frac{1}{2} \alpha C_L V_{dd}^2 f \qquad (3.8)$$

that expresses the power consumption of a CMOS inverter and, in general, of a generic CMOS gate in which internal capacitances can be neglected.

However, there can be CMOS gates in which internal nodes capacitances cannot be neglected and a more accurate analysis must be performed. In this case, voltage swings across capacitances, that can be less than V_{dd}, are responsible for a substantial charge movement between the capacitances themselves (also called *charge sharing* [63]) and/or the supply and the ground. The paths that are involved in the charge movement are dependent on the state of the gate before the transitions and the actual inputs of the gate. This implies that the power consumption of a gate is dependent on various boundary and state conditions that are not captured by a single switching activity factor. A more accurate expression for the energy consumption, taking into account internal capacitances, is:

$$dE = V_{dd}dQ_{in} - V_{gnd}dQ_{out} - (\sum_{i \in nodes} C_i V_i dV_i) \tag{3.9}$$

where dQ_{in} and dQ_{out} are, respectively, the input (from supply) and output charge quantities, C_i is the capacitance of node i and V_i is the voltage of node i, during time dt.

For dynamic CMOS circuits, charge sharing involves the movement of charges between capacitances associated to floating nodes, i.e., not connected to V_{dd} nor to the ground. In these cases, the swing across the capacitances can be less than V_{dd} but higher than 0, leading to a more complicated analysis [63].

For static CMOS circuits, voltage swings for internal nodes are always 0 or V_{dd} absolute values, depending on the state of the capacitances and the inputs to the gate. The capacitance switching always incurs in some power consumption through CMOS transistors but not in charge sharing among capacitances. Thus, whenever a capacitance C_{int} is charged to V_{dd} or discharged to 0, a $\frac{1}{2}C_{int}V_{dd}^2$ is consumed by transistors. This leads to the following expression for power consumption of static CMOS circuits [63]:

$$P = \sum_{i \in nodes} \frac{1}{2}\alpha_i C_i V_{dd}^2 f \tag{3.10}$$

Figure 3.4 shows a static CMOS NAND gate in which input A goes from 0 to 1 and input B from 1 to 0. In this situation the output remains at 0 but the internal capacitance undergoes a transition from 0 to 1. The energy consumed by the NAND gate is:

$$\int dE = \int V_{dd}dQ_{in} - \int C_{int}V_{int}dV_{int} = \frac{1}{2}C_{int}V_{dd}^2 \tag{3.11}$$

which is in accord with Equation 3.10.

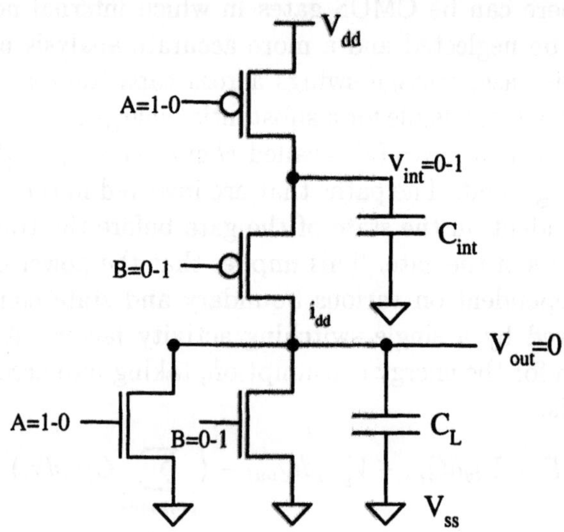

Figure 3.4. **A static CMOS NAND gate**

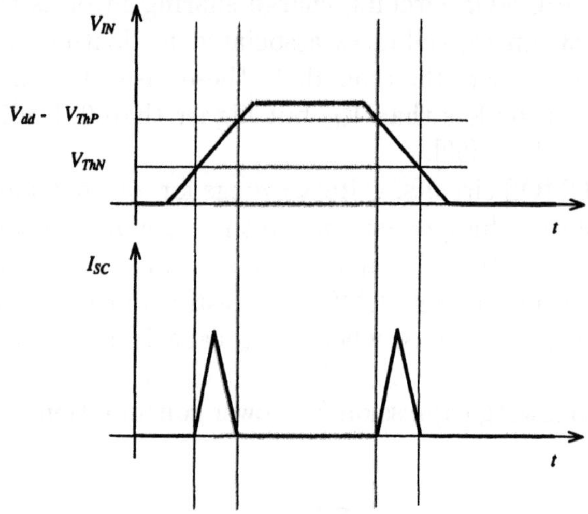

Figure 3.5. Short-circuit current generated during a transition of a CMOS inverter

1.2 Short-circuit power

Let us consider the second contribution to the average power in Equation 3.2. The short-circuit power, $P_{short-circuit}$, derives from the fact that, during a transition of a CMOS gate, both pull-up and pull-down network can simultaneously conduct, thus establishing a current flow from the supply to ground.

Figure 3.5 shows, in the upper part, the timing behavior of a switching input of a CMOS inverter (see Figure 3.1) while the lower diagram shows the corresponding current I_{SC} due to such a short-circuit.

This current is due to the fact that the signal transition time is greater than zero, thus the pull-down and the pull-up transistors are both ON for a short period of time. As a matter of fact, during a transition, there is an interval of time in which the power supply is directly connected to the ground.

Considering the threshold voltages $V_{Th,N}$ and $V_{Th,P}$ of the NMOS and PMOS devices in the inverter, the short circuit is created when the following condition is met:

$$V_{Th,N} < V_{in} < V_{dd} - V_{Th,P}. \tag{3.12}$$

Veendrick [64] first reported the expression for the short circuit power consumption of an inverter (that does not take into account load capacitance):

$$P_{short-circuit} = \frac{\beta}{12}(V_{dd} - V_{th})^3 * \frac{\tau}{T} \tag{3.13}$$

where β is the transistor strength (proportional to W/L), $V_{th} = |V_{th,P}| = V_{th,N}$, τ is the rise time of the input signal and T is the clock cycle time.

Other authors [65, 66, 67, 68] modeled the $P_{short-circuit}$ as a function of various parameters including the input capacitance, the short-channel effects and the ratio of V_{th}/V_{dd}. In particular, while the short circuit power increases by decreasing V_{th}/V_{dd}, it is shown that if V_{th}/V_{dd} is kept constant, then the ratio of the short-circuit power vs. the dynamic power remains constant even though the V_{dd} is scaled. As a matter of fact, when V_{dd} is scaled, the effect of the short-circuit power dissipation may increase and become an important part (up to 20%) of the total power dissipation of CMOS VLSIs.

1.3 Leakage power

The third component of the average power in Equation 3.2 is the power dissipated by leakage currents of MOS devices ($P_{leakage}$). The leakage current is composed of:

- a reverse saturation current (*diode leakage current*) in the diffusion regions of the PMOS and NMOS transistors and
- a sub-threshold leakage current of transistors which are nominally off.

Diode leakage is due to the voltage across the source-bulk and the drain-bulk junctions of the transistors and is the main concern for those circuits that are in stand-by mode for a large fraction of the operating time. It is usually reduced by adopting technologies with very small reverse saturation current.

Sub-threshold leakage currents are due to carrier diffusion between the source and the drain when the gate-source voltage has exceeded the weak inversion point but it is still below the threshold voltage V_{th}. In this region the MOS device behaves similarly to a bipolar transistor and thus the sub-threshold current is exponentially dependent on the gate-source voltage.

As a matter of fact, in the majority of VLSI circuits, $P_{leakage}$ can be considered as a small fraction (less than the 10%) of the total power consumption.

2. Power estimation techniques

By *power estimation*, we generally refer to the problem of estimating the average power dissipation of a digital circuit. Direct power estimation can be done at every abstraction level and could provide a feed-back on the optimization phase of the design enabling the exploration of multiple design alternatives. The effectiveness of such type of exploration depends obviously on the degree of accuracy of the power characterization. For example, a power optimization that leads to 10% of power savings on a particular model is effective if the power model is characterized by 1% of error in the accuracy, but could be insignificant if the model is characterized by an average 15% of error in the accuracy.

Power estimation techniques can be divided in two classes: dynamic and static [69]. Dynamic techniques explicitly simulate the circuit under a "typical" input stream and are typically characterized by a very high number of test vectors. These techniques usually provide a high level of accuracy, but are characterized by a very high simulation time. Static techniques rely on probabilistic information about the input stream (e.g., switching activity of the input signals, temporal correlations, etc.) in order to estimate the internal switching activity of the target circuit.

In the following subsections, an overview of the most important static and dynamic techniques available for power estimation at the different abstraction levels is provided, highlighting their accuracy and performance.

2.1 Transistor-level power estimation

Power estimation at the transistor level can be performed by keeping track of the current drawn from the power supply during a transistor level simulation. Very few approximations are made and typically all components of power consumption - both static and dynamic - are analyzed. In fact, the main limitation on accuracy is due more to an incomplete knowledge and modeling of parasitics and interconnect than to inaccuracies in the device models. This problem can be solved when a circuit layout is available and the wiring parasitics can be extracted.

SPICE, CAzM, Powermill and IRSIM can be used to perform power estimation at the transistor level. In particular, IRSIM has been extended to IRSIM-CAP [70] in order to improve the measurement of glitching power, as well as incremental power measurement. IRSIM-CAP models power consumption based on a "three-level quantization scheme" where the voltage can assume three values (GND, $V_{dd}/2$ and V_{dd}). Three-level quantization scheme has shown to be more accurate than the two-level rail-to-rail model by preserving original circuit simulation performance.

The power estimates of IRSIM-CAP are in fact reasonably close to those that can be derived by measuring currents with a SPICE simulator, with an error less than 20% and a speed up of about 580x. LDS [71] is another switch-level simulator that tries to estimate also short-circuit and static power by using some analytical approximations triggered by signal slopes.

2.2 Gate-level power estimation

Gate-level power estimation can be performed either *probabilistically* or by simulation.

Probabilistic gate level power estimation is focused on switching activity estimation by means of some a-priori information about the input stream.

The power consumption of a specific gate, can be computed by estimating the switching activity of its output and applying equation 3.8. The output switching activity is computed by estimating the signal probability of the output p_f and then computing the transition probability $2p_f(1-p_f)$ [72].

The output signal probability is usually computed by using the canonical expression of the gate functionality; in the case of sum of products ($f(x_1, \ldots, x_n) = \sum_j \prod_k s_k, s_k = x_k$ or $\overline{x}_k$), we have that the signal probability of f is expressed as:

$$p_f = \sum_j \prod_k p_{s_k}. \tag{3.14}$$

As an example, let us consider a two input OR gate whose input signal probabilities are p_1 and p_2. By assuming both spatial and temporal independence [73], the signal probability of the output (p_o) can be com-

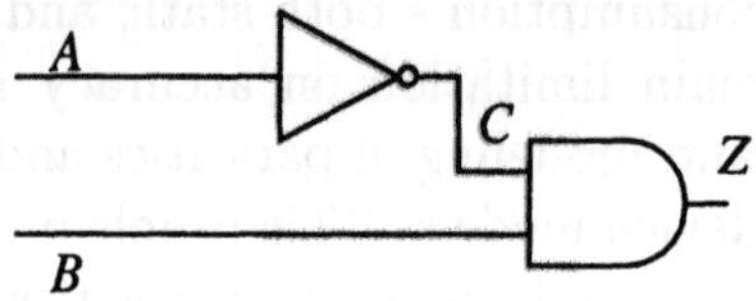

Figure 3.6. **Logic circuit without reconvergent fanout**

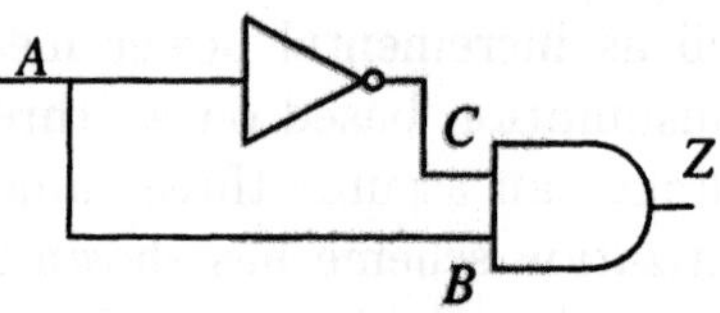

Figure 3.7. **Logic circuit with reconvergent fanout**

puted as the inverse of the probability that the two inputs are at the logic level 0:

$$p_o = 1 - (1 - p_1)(1 - p_2) \tag{3.15}$$

The switching activity of the OR gate is then computed as the probability that the output makes a transition $0 \rightarrow 1$ or viceversa between two clock cycles and can be computed in this way:

$$\alpha = p_o^{0\rightarrow 1} + p_o^{1\rightarrow 0} = p_o(1 - p_o) + (1 - p_o)p_o. \tag{3.16}$$

To estimate the power consumption of a set of gates, in which there are no feedback and input signals are mutually independent, the output probabilities can be used as input probabilities of the next level of logic to simulate the switching activity propagation.

For example, consider the logic network in Figure 3.6. If the two primary inputs A and B are uncorrelated and are uniformly distributed ($p_A^1 = \frac{1}{2}$, $p_B^1 = \frac{1}{2}$), then the probability that the node Z transitions from 0 to 1 is the following:

$$p_Z^{0\rightarrow 1} = (1 - p_C^1 p_B^1) p_C^1 p_B^1 = (1 - \frac{1}{4})\frac{1}{4} = \frac{3}{16} \tag{3.17}$$

given that $p_C^1 = p_A^1$. In this case the switching activity associated with Z is $\alpha_Z = p_Z^{1\rightarrow 0} + p_Z^{1\rightarrow 0} = \frac{3}{8}$.

In the case of reconvergent fanout, however, the input signals to a gate cannot be considered as mutually independent. For example, let us consider the logic network in Figure 3.7 where the inputs of the AND gate (C and B) are interdependent, as B and C are also function of A.

The presented approach to compute signal activity for Z by assuming B and C independent would have given the same results of the previous example ($\alpha_Z = \frac{3}{8}$) while, as a matter of fact, the node Z never switches.

To solve these problems, signal inter-dependencies must be taken into account. Referring to the example of Figure 3.7, the signal Z is 1 if and only if B and C are equal to 1. Thus, the signal probability of Z can be conditioned to the probability that B is 1 given that C is 1:

$$p_Z = p(B = 1 \wedge C = 1) \tag{3.18}$$

If B and C are independent, $p(B = 1 \wedge C = 1)$ reduces to $p_B p_C$, while if they are dependent:

$$p_Z = p(B = 1 \wedge C = 1) = p(C = 1 | B = 1) p_B. \tag{3.19}$$

In this case, conditional probabilities (such as $p(B = 1 \wedge C = 1)$) must be computed by taking into account the circuit functionality.

An alternative approach is to compute the signal probability of node Z by applying Eq. 3.14. Other authors [63] perform signal probability estimation by using Binary Decision Diagrams (BDD) [74] and decomposing boolean function in *cofactors* and estimating the probability of the cofactors.

In presence of correlated signals, the switching activity for a complex boolean circuit can be also analyzed by using the notion of boolean difference of a function f with respect to one of its independent variables x_j [63]:

$$\frac{\partial f}{\partial x_j} = f|_{x_j=1} \oplus f|_{x_j=0}. \tag{3.20}$$

The boolean difference is used to account for the function sensibility to the changes on x_j. In fact, in case inputs are not subject to a transition simultaneously, the switching activity of f can be estimated as:

$$\alpha(f) = \sum_{j \in inputs} P(\frac{\partial f}{\partial x_j}) \alpha(x_j) \tag{3.21}$$

where $P(\ldots)$ is the signal probability. The authors of [75] extend Equation 3.21 to work with simultaneous transitions on the inputs by means of *generalized boolean differences*.

Other autors [76] use zero-delay hypothesis and lag-one Markov chains to model spatiotemporal correlations among the primary inputs and internal lines of a circuit and they propose efficient heuristics for probabilities and correlation coefficients estimation.

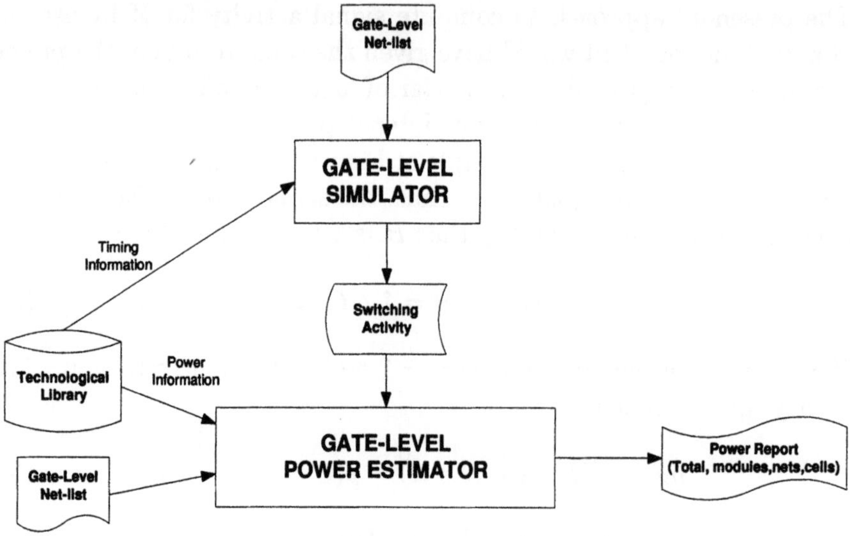

Figure 3.8. Gate-level power estimation flow

2.2.1 Gate-level simulation based estimation

Gate-level simulation-based power analysis is based upon the measurement of the switching activity in every net of the circuit during a gate-level simulation. The switching activity is combined by a power estimator tool (such as Synopsis DesignPower) with information from the technological library to compute total power dissipation (see Figure 3.8). Such type of tools subdivide the power dissipation of a gate (i.e., a cell) in three major components:

- $P_{switching}$ (is associated with the output switching activity and depends on the wire-load model)
- $P_{internal}$ (also called *short-circuit*, depends on the internal nodes capacitance switching and short-circuit currents)
- $P_{leakage}$ (associated with the leakage current flowing in the gate).

$P_{internal}$ is a function of the cell's state, the input vector that is currently applied, the slope of that input vector and the load of the output pin and it is stored in the cell structure of the technology library by means of lookup tables. This type of power characterization is also called *state-dependent path-dependent* (SDPD). A high-quality library and a good simulation model for interconnect, glitching, and gate delays can produce power estimates extremely accurate that can be used as reference for estimates at higher levels of abstraction. However high accuracy requires tipically a large set of input vectors and consequently a long simulation time. As a consequence, a number of stream compaction and synthesis technique have been introduced in the literature. These techniques try to reproduce a new input signal with either the same spatial and temporal correlations [77] or the same spectral characteristics [78] of the original stream.

2.3 RT-level power estimation

Power estimation at register-transfer (RT)-level [79] is fundamental when the main concern is shortening the design cycle. In this technique, transistor-level or gate-level simulations of RT-level modules under their respective input sequence are replaced by power macro-model evaluation. Macro-models consist of capacitance models for circuit modules and switching activity profiles for data and/or control signals. The macro-model for the components can be parameterized in terms of the input bus width, the internal organization/architecture of the component, and the supply voltage level.

The evaluation of the model consists of extracting the actual parameter values for the macro-model equation either from static analysis of the circuit structure and functionality or by performing a behavioral simulation of the circuit. In the latter case, a power co-simulator linked with a standard RT-level simulator can be used to collect input data statistics for various RT-level modules in the design.

The power factor approximation (PFA) technique [80] uses a weighting factor, determined through experiments, to model the average power consumed by a specific module, over a range of designs, per input change. The weakness of this technique is that it does not account for the data dependency of the power dissipation.

Dual bit type model is another macro-model for high-level modules, proposed by Landman *et al.* [81], exploits the fact that, in the data path, switching activities of high-order bits (i.e., sign bits) depend on the temporal correlation of data while lower order bits behave randomly. Thus a module is completely characterized by its capacitance models in the MSB and LSB regions. The break-point between the two regions is determined based on the signal statistics collected from simulation runs. Landman proposed also the activity-based control (ABC) model [82] to estimate the power consumption of random-logic controllers.

An input-output model has been proposed by Gupta *et al.* [83] to capture the dependence of power dissipation in a combinational logic circuit on the average input signal probability, the average switching activity of the input lines, and the average (zero-delay) switching activity of the output lines. Parameter extraction is obtained from a functional simulation of the circuit that is usually very fast. The authors also present an automatic macro-model construction procedure based on random sampling principles.

One common feature of the above macro-model techniques is that, they only provide information about average power consumption over a relatively large number of clock cycles. The above techniques, which are suitable for estimating the average-power dissipation, are referred to as cumulative power macro-models. In some applications, however, estimation of average power is not sufficient and a cycle-based power modeling is needed.

Wu *et al.* [84] propose a regression-based methodology to build cycle-accurate models. The methodology is based on statistical sampling for the training set design and regression analysis combined with an appropriate statistical test. The test is a sort of principal component analysis that identifies the most power-critical variables. The statistical framework enables also to compute the confidence level for the predicted power values.

Qiu *et al.* [85] extended the previous model to capture "important" up to three order of spatial correlations at the circuit inputs. Note that here the equation form and variables used for each module are unique to that module type. This power macro-model has been able to predict the module power with a typical error of 5-10% for the average power and 10-20% for the cycle power.

Other authors [86] propose to evaluate the macro-model equation not at each clock cycle but only within random samples of the input trace. Experimental results have shown that this technique has an average efficiency improvement of 50X over the full cycle-by-cycle macro-modeling, with an average error of 1%.

In [87], a fast cycle-based simulation is used until there are significative changes in the local mean value of the input stream. At this point, an accurate power simulation (transistor or gate-level) is automatically started for limited interval of simulation time. Experimental results have shown a simulation speedup of more than 20X over transistor level simulation.

In [88], the authors present an efficient macromodeling technique for RTL power estimation, based only on word and bit level switching information of the module inputs. The paper shows that the developed models reduce the estimation error compared to the Hamming-distance model at least by 64

In [89], the authors propose a new power macromodel for usage in the context of register-transfer level (RTL) power estimation. The model is suitable for reconfigurable, synthesizable, soft macros be-cause it is parameterized with respect to the input data size (i.e., bit width) and can also be automatically scaled with respect to different technology libraries and/or synthesis options.

2.4 Microarchitecture-level power estimation

Microarchitectural power estimation techniques are less accurate, but faster than gate-level and RT-level estimation ones. These techniques [90, 91, 92] require a functional model of the processor at the level of the major components (ALU, register file, etc.) along with a model (in general, a look up table) of the power consumption for each module. These approaches use as stimulus a trace of executed software instructions (usually derived from a functional simulation) and consider the contribution of each processor module activated by a single instruction to derive the total energy consumption per cycle.

The most recent works in the area of micro-architectural power estimation address high-end processors with complex microarchitectures [93, 92, 94]. Similarly to instruction-level power analysis, these estimators are based on an Instruction Set Simulator, but they feature a detailed micro-architectural model of the processor, with dedicated power models for its main functional units (such as execution units, pre-fetch buffers, register files, on-chip caches, etc.). Micro-architectural power modelling is particularly attractive for processor architects, aiming at exploring the design space parameters of processor and cache organizations. Absolute accuracy with respect to the final gate-level or transistor-level implementation is not a must for this type of models, and it is hardly achievable, because the detailed circuit design is usually completed after the micro-architectural exploration. The aim of power modelling is to achieve a good level of relative accuracy when comparing different architectures over a wide range of hardware configurations.

Wattch [94] is a framework for analyzing and optimizing microprocessor power dissipation at the architecture-level. Experimental results have shown that Wattch is 1000X faster than existing layout-level power modeling tools with an accuracy within 10% verified on industrial designs. Wattch has been built by integrating the power model of common structures within Simplescalar, a configurable and efficient architectural simulator. In this way, Wattch allows the designer a very fast design space exploration from the point of view of both power dissipation and performance.

SimplePower [93] is an energy simulator based on transition-sensitive energy models, and it is an exececution-driven cycle-accurate microarchitectural simulator of a superscalar architecture equivalent to SimpleScalar. SimplePower features technology dependent power estimation capabilities for the data-path, the caches and the bus power consumption and enables efficient design space exploration.

Some recent approaches [95] exploit embedded hardware monitors (e.g., processor performance counters) to investigate the energy usage patterns of individual threads. The information about active hardware units (e.g., integer/floating-point unit, cache/memory interface) is gathered by event counters to establish a thread-specific energy accounting. In fact, it is shown that a correlation exists between events and energy values.

In [96] a power-performance modeling toolkit for PowerPC processors has been developed. The power estimation engine is based on a fast, cycle-accurate, parameterized research simulator and a slower, pre-RTL reference model that simulates high-end machines in full, latch-accurate detail. Energy characterizations are derived from real, circuit-level power simulation data and combined with microarchitecture-level parameters to provide energy estimates.

In [92], a tool set for power estimation around the existing, research and production-level simulators for PowerPC processors is presented. The tool set is based on the plug-in of the energy models into a workload driven cycle simulator. In each simulated cycle, the activated microarchitecture-level units are known from the simulation state and, depending on the particular workload, a fraction of the processor units and queue/buffer/bus resources are working. Accurate energy models are then used to estimate the total energy spent on a unit basis as well as for the whole processor.

In [97], the authors propose AccuPower, a true hardware level and cycle level microarchitectural simulator. Energy dissipation coefficients are taken from SPICE measurements of actual CMOS layouts of critical datapath components. Transition counts are obtained at the level of bits within data and instruction streams, at the level of registers, or at the level of larger building blocks (such as caches, issue queue, reorder buffer, function units).

2.5 Instruction-level power estimation

This type of software power estimation is based on a functional simulation of the instruction set of the processor. During the instruction set simulation, an instruction-level model associates a black-box cost to each instruction, by considering also circuit state effects, pipeline stalls and cache misses produced during its execution.

Research on instruction-level power analysis (*ILPA*, for brevity) has started only recently: thus, only few proposals have been made on this subject, that yet lacks of a rigorous approach.

The major contributions found in literature are empirical approaches based on physical measurements of the current drawn by the processor during the execution of embedded software routines. Tiwari *et al.* [98, 99, 100] proposed to estimate the energy consumption of a processor by an instruction-level model that assigns a given power cost (namely the *base energy cost*) to each single instruction of the instruction set. However, during the execution of a software routine, certain *inter-instruction effects* occur, whose power cost is not taken into account if only the base energy costs are considered. The first type of inter-instruction effects (namely the *circuit state overhead*) is associated with the fact that the cost of a pair of instructions is always greater than the base cost of each instruction in the pair. The remaining inter-instruction effects are related to resource constraints that can lead to stalls (such as pipeline stalls and write buffer stalls) and cache misses which imply a penalty in terms of energy.

Globally, the model in [98, 99] expresses the average energy as:

$$E = \sum_{i}(B_i \times N_i) + \sum_{i,j}(O_{i,j} \times N_{i,j}) + \sum_{k} E_k \tag{3.22}$$

where E is the total energy associated with the execution of a given program, which is computed as the sum of three components. The first component is the sum of the energy base cost B_i of each instruction i multiplied by the number of times N_i that the instruction is executed. The second component is the sum of the additional energy cost $O_{i,j}$ due to the elementary instruction sequence (i, j) multiplied by the number $N_{i,j}$ of executions of each sequence; such a contribution is evaluated for all instruction pairs. The third contribution E_k takes into account any additional energy penalty due to resource constraints that can lead to pipeline stalls or cache misses.

The basic idea of this approach is to measure B_i as the current drawn by the processor as it repeatedly executes a loop whose body contains a sequence of identical instructions i. The term $O_{i,j}$ takes into account the fact that the measured cost of a pair of instructions has been always observed to be greater than the sum of the base costs of each single instruction. These terms must be stored in a matrix whose size grows with the square of the number of instructions in the ISA.

This turns out to be impractical for CISC processors with several hundreds of instructions, and the storage problem becomes exponentially complex when considering VLIW architectures, in which long instructions are composed of arbitrary combinations of parallel operations. A possible solution is presented in [99], where the authors propose to reduce the spatial complexity of the problem by clustering instructions based on their cost. The results of the analysis motivated several software-level power optimization techniques, such as instruction re-ordering to exploit the power characteristics of each instruction.

The instruction-level power model proposed in [101] considers an average instruction energy equal for all instructions in the ISA. More specifically, this model is based on the observation that, for a certain class of processors, the *energy per instruction* is characterized by a very small variance. In this case, note that while in complex architectures like CISCs, the heavy use of caches and microcode ROMs produces a relevant increase of the average instruction energy and a reduction of the energy variance, this is not always true for RISC and VLIW architectures for which these effects can become particularly evident.

Another attempt to perform instruction level power estimation is proposed in [102], where the authors consider several inter-instruction effects as well as statistics of data values used within the processor. Although the developed power model is quite accurate, it lacks general applicability, being developed only for a specific embedded processor.

In [103], inter-instruction effects are measured by considering only the additional energy consumption observed when a generic instruction is executed after a NOP (the proposed power model is also called the *NOP model*). As a matter of fact, the model reduces the spatial complexity proper of instruction-level power models.

In [104], the authors present an efficient and accurate high-level software energy estimation methodology using the concept of characterization-based macro-modeling. In characterization-based macro-modeling, a function or sub-routine is characterized using an accurate lower-level energy model of the target processor, to construct a macro-model that relates the energy consumed in the function under consideration to various parameters that can be easily observed or calculated from a high-level program.

2.6 System-level power estimation

Typically, system-level power estimation considers all the hardware modules composing the system such as hard disks, I/O or network interfaces. This type of power estimation can be less accurate than instruction-level power estimation but much more faster; this is useful, for example, during the initial design of a complex system when many details of the implementation are not defined or are subject to change. System-level power estimation can be performed also when the an ISS of the target system is available together with the entire set of power models for all the hardware modules of the system.

In recent years, few approaches to system-level power estimation have been studied [105, 106, 107]. These approaches consider a constant power dissipation value associated with each component of the system and compute the total power dissipation by simply summing all the constant contributions. The main limitation of these approaches is that they don't take into consideration the system workload and the possible interactions between the various components.

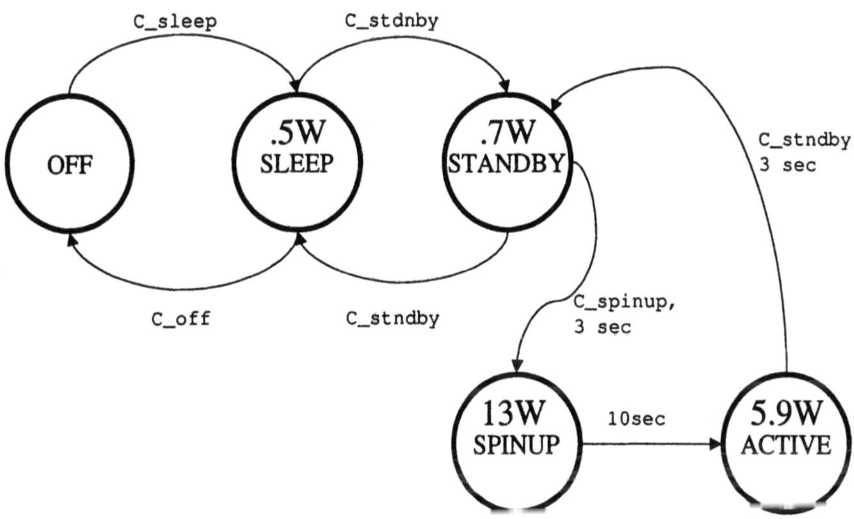

Figure 3.9. A power state machine for a disk drive

Another approach proposed by Benini *et al.* [108] tries to overcome this limitation by using a more complex system model where the workload computation is automated by specifying the interaction among components. In this model each resource is described by a simple state machine, containing only information relevant to its power behavior (see Figure 3.9 for an example of power state machine for an hard-disk). At any given point in time, each resource is in a specific power state, and consumes a specified amount of power. State transitions are executed by specific commands on the resource and imply a performance penalty. In this model, resources can be also *utilization-dependent* and, in this case, they have a corresponding activity level that is the fraction of the peak power that is consumed for that level of usage. The activity level of a resource during operation can be modified without changing the resources power state. Finally the concept of *environment* is used for representing a system workload. It consists of a series of external (user) requests over time. These requests must be tracked over a representative time period for the overall model to have validity.

Such type of power estimation is useful to develop dynamic (or *stochastic*) power management at the system-level, described in Section 6.

Chapter 4

INSTRUCTION-LEVEL POWER ESTIMATION FOR VLIW PROCESSOR CORES

In this chapter, an instruction-level energy model for the data-path of VLIW pipelined processors is proposed. The model provides accurate power consumption estimates during an instruction-level simulation and it can be also used at compile time to drive a power-oriented scheduling. The analytical model we propose takes into account some software-level parameters (such as instruction ordering, pipeline stall probability and instruction cache miss probability) as well as some other parameters at the microarchitectural-level (such as pipeline stage power consumption per instruction). Globally, the model represents an efficient pipeline-aware instruction-level power estimation framework, whose accuracy is very close to those given by RT or gate-level simulations. The general problem of the power characterization at the instruction-level of a K-issue VLIW processor has a complexity of $O(N^{2K})$, where N is the number of operations in the ISA and K is the number of parallel instructions composing the very long instruction. The proposed model reduces the complexity of the characterization problem to $O(K \times N^2)$. The model has been used to characterize a 4-issue VLIW custom core with a 6-stage pipeline. The accuracy and efficiency of the proposed model has been compared with respect to the energy estimates obtained by the gate-level simulation. A set of experimental results carried out on standard embedded multimedia benchmarks are reported in the chapter, showing an average error in accuracy of 4.8% of the instruction-level estimation engine with respect to the gate-level engine. The reported experimental results shown also an average simulation speed-up of the instruction-level power estimation engine with respect to the gate-level engine of four orders of magnitude.

V. Zaccaria et al. (eds.),
Power Estimation and Optimization Methodologies for VLIW - based Embedded Systems, 41–65.

1. Introduction

The low power issue represents one of the most important requirements for many classes of embedded systems [109]. Efficient techniques to access power estimation at the highest levels of abstraction is of primary importance for successful system-level design [110, 79]. Starting from the pioneer work of Landman and Rabaey [81] on the high-level power modeling of data path modules, many research works appeared recently in literature to afford the problem of increasing the level of abstraction of the power estimation process. Some of these works have addressed the problem of power estimation for high performance microprocessors, where pipelining and instruction-level parallelism must be considered from the early phases of the design process. However, many open issues must be investigated in this area, and an efficient and accurate framework for power estimation at the system-level has not been developed yet.

Main goal of our work is to define a power-aware exploration methodology at the system-level suitable for VLIW (Very Long Instruction Word) embedded processor cores. In general, a VLIW processor is a pipelined ILP (Instruction Level Parallelism) processor that can execute, in each clock cycle, a set of explicitly parallel *operations*; this set of operations is statically scheduled to form a *Very Long Instruction Word.* The approach described in this chapter is an extension of the work we previously proposed in [111, 112, 113], addressing an instruction-level power model to estimate the software energy consumption associated with the pipelined VLIW core. The model is quite general and parametric and it enables the exploration of the power budget by considering different solutions both at the compiler-level and at the architectural-level, given a specific application.

1.1 Overview of the Work

In general, the instruction-level power analysis does not give any information on the instantaneous causes of power consumption in the processor core, which is seen as a black-box model. Moreover, this approach does not consider the power contributions due to the single operations that are active in each pipeline stage. These two issues can impact the capability to execute a realistic run-time power profiling of a software application, because the overlapping of instructions within the pipeline is discarded. Furthermore, this approach can exclude some specific, compiler-level optimization techniques, such as defining instruction schedules whose power consumption is always under a given upperbound or with reduced variation rates (to optimize the battery usage).

When considering VLIW processors, traditional instruction-level power modeling approaches imply new challenges, since the number of long instructions that can be created by combining N operations into a K-wide VLIW is N^K. It becomes evident that traditional instruction-level approaches, requiring the storage of the energy parameters for each pair of instructions, are not applicable to VLIW cores.

To address these open issues, the main focus of this work consists of defining an energy model for VLIW architectures combining the efficiency and the flexibility of instruction-level models with the accuracy of microarchitectural level ones. The main goals are, from one side, to give accurate estimates on the energy consumed by the processor core during an instruction-level simulation of a VLIW embedded application and, from the other side, to provide to the compiler a fine grained energy model of the instructions overlapped in the pipeline. The long term goal of our work is to support the power optimization techniques from both the compilation-level and software-level perspectives.

Our analysis mainly addresses the data-path of VLIW cores. The energy model is based on an analytical decomposition of the global energy associated with a software instruction into its contributions due to the microarchitectural components (i.e. the active processor function units and the pipeline stages). These microarchitectural contributions are associated with the parallel operations executed along the pipeline stages combined with other instruction-level contributions (i.e. the probability and the latency per instruction of a miss in the memory hierarchy). When using a traditional black-box instruction-level model, run-time accuracy cannot be achieved because black-box model ignores the overlapping of the instructions in the pipeline.

A traditional energy model can be used to evaluate, within an ISS (Instruction Set Simulator), the energy for each given clock cycle. This situation is pointed out in Figure 4.1-a, where we can note how the traditional model associates with each clock cycle the instruction that is being currently fetched from the cache, neglecting (totally or mostly) the effective presence in the pipeline of other instructions and by modeling cache misses as a fixed power cost to be added to the current instruction. Figure 4.1-b describes the situation when considering our pipeline-aware model, where the power consumption is modeled on a *per-cycle-basis* by splitting the energy of each instruction in the energy contributions of the corresponding stages and by summing over the several contributions at any given clock cycle. In particular, in the reported example, our pipeline-aware power model considers a pipeline composed of six stages with a uniform decomposition of the energy of each instruction over the single pipeline stages. Of course, such type of models can be used within

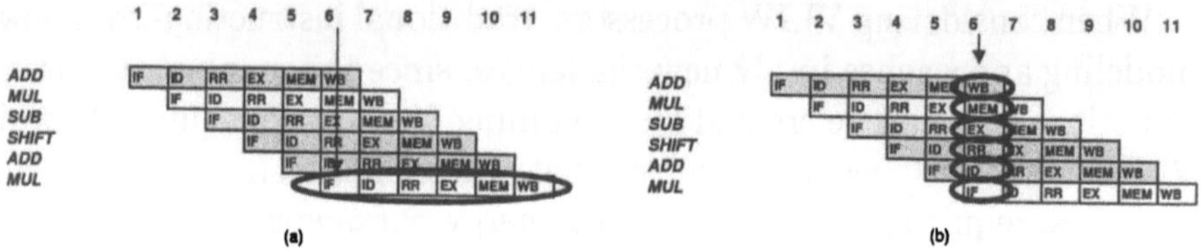

Figure 4.1. Traditional (a) and pipeline-aware (b) instruction-level energy models

an ISS, that considers the pipeline structure of the processor (i.e. cycle-based ISS). Obviously, the value of the *average* energy per cycle for a given instruction sequence is identical for both types of models, as shown in Figure 4.1.

Compared to a traditional model, our pipeline-aware power model is much closer to the real pipeline behavior, since it effectively considers the effects of the instructions overlapping within the pipeline. An example of comparison of the results derived from the application of a traditional instruction-level power model and the pipeline-aware power model to the same sequence of instructions is shown in Figure 4.2. The reported sequence, derived from a real application, is characterized by two bursts of instructions separated by a sequence of NOPs inserted by the compiler (where the energy values are in the range from 1490 [pJ] to 1919 [pJ]). As expected, even if the average energy per cycle is equal for both models, the energy behavior for each clock cycle is quite different. In each clock cycle, the traditional model considers the energy contribution associated with the instruction in the fetch stage. The pipeline-aware model has a more precise insight on the different instructions present in the pipeline at any given clock cycle. This effect becomes particularly evident during the filling up (flushing) phases of the pipeline after (before) a sequence of NOP instructions. Furthermore, the pipeline-aware model reproduces a smoother energy behavior with respect to the traditional model. The traditional model points out some peaks, not found in the real-world energy behavior, outlining the differences in terms of energy cost for each instruction.

Globally, the pipeline-aware model can be used to add value either during an instruction-level simulation, to realistically model the energy behavior of a processor, or during a power-aware compiler that can modify power figures (i.e., spikes, upper-bounds and rates of variation) by means of instruction scheduling. An application example of such type of compilation can be the re-scheduling phase to reduce the variation rates in the energy use to extend the battery life. Another type of application could be a more power-aware instruction scrambling algorithm that

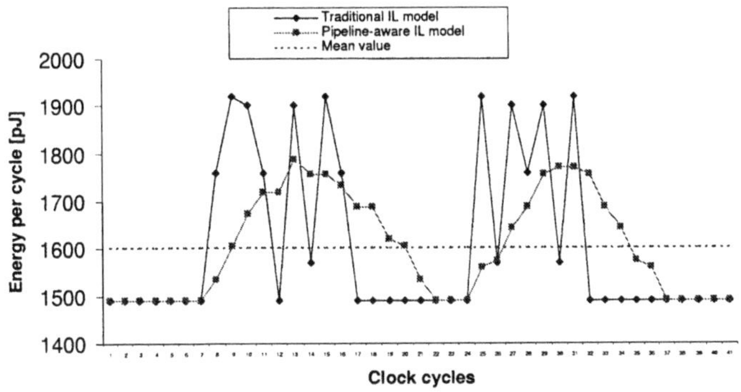

Figure 4.2. Comparison between traditional and pipeline-aware instruction-level energy models.

tries to make uniform the energy consumption for a given cryptography application, to avoid security attacks, based on the observation of the power behavior.

Our pipeline-aware energy model is based on the consideration that the execution of an instruction can be projected on two axes: a *temporal* one and a *spatial* one.

From the *temporal projection* perspective, we trace the execution of a very long instruction through the pipeline stages. The model exploits the temporal projection to evaluate the energy associated with an instruction as the sum of the energy costs due to each pipeline stage concerned in its execution. From this perspective, we can say that the energy is *temporal additive.*

The *spatial projection* draws the use of the resources of the processors within each single pipeline stage. The model exploits the spatial projection to split the energy of a pipeline stage into the sum of the energy costs due to each operation involved in the execution of the current and the previous long instruction. To this purpose, we assume that the mutual influence between two different operations of the same very long instruction can generate an irrelevant effect with respect to the mutual influence between operations of two successive very long instructions. This feature of the model, indicated as the *spatial additive property* of the energy of a VLIW processor, has been experimentally verified in our set of experiments and it is of fundamental importance to deal with the complexity of the characterization problem of the instruction-level power model for VLIW cores.

Based on both temporal and spatial additive properties, the problem complexity has been reduced from $O(N^{2K})$ to $O(K \times N^2)$, where K is the number of parallel operations in the very long instruction and N is the number of operations in the ISA. Generally, a more accurate estimation than a black-box model is reached with a reasonable computational complexity.

Furthermore, the main difference of our approach with respect to *ILPA*-based approaches consists of the fact that our method provides a better view on the power bottlenecks during the software execution, since it is based on a detailed microarchitectural model of the processor core and it analyzes the energy behavior on a per-cycle-basis. In particular, the proposed analytical model defines a precise mapping between long instructions and microarchitectural functional units used during the instruction execution. This instruction-to-unit mapping is used to assess RT-level energy estimates for each unit, to be combined with stall and latency energy contributions to derive instruction-level power estimates.

To conclude, aim of our work is the definition of an accurate power estimation framework for the end-user to estimate the power consumption associated with an embedded software code. Furthermore, we can envisage the use of the proposed framework for system-level power exploration during the compilation phase (through a power-oriented scheduler) as well as during the microarchitectural optimization phase of the processor design.

The chapter is organized as follows. The overall architecture of the target system is described in Section 2. Section 3 introduces the general instruction-level energy estimation model for pipelined VLIW architectures. Some experimental results derived from the application of our model to a 4-issue VLIW in-house architecture are reported in Section 4. Finally, Section 5 concludes the chapter by outlining some research directions originated from this work.

2. System Architecture

In this section, we describe the processor architecture, the ISA (Instruction Set Architecture), and the processor requirements for the target system.

2.1 Target Processor Architecture

The target processor architecture is based on a load/store VLIW core processor (see Figure 4.3). In general, a VLIW processor is a pipelined CPU that can execute, in each clock cycle, a set of explicitly parallel *operations*. This set of operations is indicated as *very long instruction word*. In this chapter, we abbreviate the term very long instruction as *instruction*, while maintaining the use of the term *operation* to identify the single components of an instruction. The position assigned to an

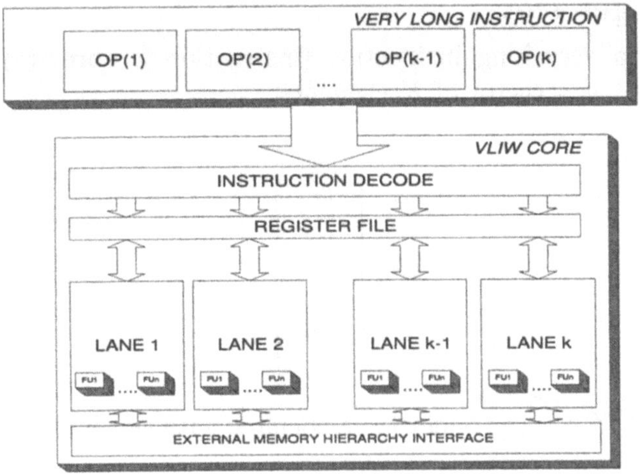

Figure 4.3. Architecture of the target VLIW pipelined core

operation within an instruction is called *slot*, that usually defines the execution path (also called *lane*) of the operation within the VLIW processor. A K-issue VLIW processor can fetch an instruction composed of one to K parallel operations and it executes them on K parallel lanes (see Figure 4.3). In a pipelined VLIW, we can assume that each lane executes a single operation in a pipelined fashion. Generally, the decode stage decodes an instruction in several control words associated with each operation and these words are then used to control the pipeline stages of each lane. Ideally, even the decode stage can be decomposed into lanes that decode each single operation of the long instruction.

We assume the target VLIW architecture does not support predication, since this mechanism could waste power, while executing instructions that will not be committed back to the registers. We further assume that the target system architecture includes separate instruction and data caches.

When a miss in the instruction cache occurs, the instruction cache propagates to the core an 'IC-MISS' signal, that is used, within the core, to 'freeze' the inputs of the major part of the pipeline registers, avoiding useless switching activity. The behavior is similar to the execution of NOP instructions inserted in the code at compile time. The only difference between the two cases is that explicit NOP instructions generate additional switching activity passing through the pipeline registers. When a miss in the data cache occurs, the core gates the clock to all the pipeline registers until the miss is served.

2.2 Target ISA

We define a very long instruction (instruction for brevity) $\mathbf{w}$ of a K-issue VLIW as a vector of K elements:

$$\mathbf{w} = \begin{bmatrix} w^1 \\ \dots \\ w^k \\ \dots \\ w^K \end{bmatrix} \tag{4.1}$$

where $w^k \in ISA$ is the operation associated with slot k of the instruction $\mathbf{w}$.

Considering power issues, it has been observed in [99] that the operations most commonly used in a target application can be categorized into classes, such that the operations belonging to a given class are characterized by similar power costs. The basic assumption is that operations with similar functionality activate similar parts of the processor core, so they have similar power behavior. The spatial additive property (described in more detail in Section III) means that the several lanes will be considered mutually independent with respect to power, so that power requirements of single operations can be analyzed without considering the relationships with other operations in the same instruction.

In this work, we suppose that two operations in the ISA as "not equal" if they differ either in terms of basic functionality (i.e., opcode) or in terms of addressing mode (immediate, register, indirect, etc.). Even without considering data differences (either in terms of register names or immediate values), the number of pairs of operations to be considered could become too large to be characterized based on a transistor-level or even a gate-level simulation approach. For example, in the case of an ISA composed of 70 operations, with several possible addressing modes for each operation, it is necessary to characterize 6126 pairs of operations to consider inter-instruction effects.

To reduce the number of experiments during the characterization phase, we applied the cluster analysis proposed in [114] to the operations of the ISA for VLIW architectures. The cluster analysis consists of grouping in the same cluster the operations with similar power costs. The power cost of an operation has been defined as the power consumed by the processor when it executes only that operation.

In the rest of the chapter, we assume that the ISA of the target VLIW processor is clustered into classes of operations that have similar power behavior.

2.3 Target Processor Requirements

The proposed approach requires a gate-level or (at most) RT-level description of the target processor, so that accurate energy values can be derived for the characterization of the model. To store the information needed for the proposed energy model at run-time , it is necessary to make use of an ISS suitably instrumented.

3. VLIW Energy Model

This chapter represents an extension of the work we previously proposed in [111, 112, 113], focusing on an instruction-level energy estimation model for the data-path of VLIW embedded cores. The model is based on an analytical energy model that considers the processor microarchitecture. The energy of a very long instruction is considered as the sum of a set of energy contributions due to the single pipeline stages involved in the instruction execution and to the single operations composing the instruction.

Generally, in a VLIW processor, instructions are composed of one or more explicitly parallel operations. After an instruction is fetched from the I-cache, each operation is dispatched to a specific functional unit (see Figure 4.3). In contrast with superscalar processors, VLIW processors do not check for any data or control dependencies, since the flow of operations without any types of *intra* or *inter*-bundle dependency is statically guaranteed by the compiler.

As already noted in [103], the main issue of an instruction-level power model for ILP processors derives from the complexity due to the spatial and temporal correlations of the instructions in the execution trace. As noted before, to consider each inter-instruction effect we need an n-dimensional array, where n depends on the type of inter-instruction effects. For example, if we only need to take into account the inter-instruction effect between adjacent instructions, it is necessary to store this information in a bi-dimensional array, whose size is proportional to the number of instructions in the ISA. In the case of a RISC or CISC machine, some authors proposed to reduce the problem complexity by clustering instructions [99], thus reducing the length of the parameters array.

For K-issue VLIW architectures, the power model must consider all the possible combinations of operations in an instruction, thus the problem complexity grows exponentially with K and the number of operations in the ISA. Instruction clustering can be much more effective to reduce the problem complexity, by trading off power estimation accuracy and problem complexity, but the problem complexity can be further simplified as shown in the rest of the chapter.

Let us consider a sequence $\mathcal{W}$ composed of N very long instructions:

$$\mathcal{W} =< \mathbf{w_1}, \ldots, \mathbf{w_{n-1}}, \mathbf{w_n}, \ldots, \mathbf{w_N} > \tag{4.2}$$

where $\mathbf{w_n}$ is the n-th instruction (see Equation 4.1) of the sequence.

In this work, we assume the energy associated with $\mathbf{w_n}$ dependent on the properties of $\mathbf{w_n}$ in terms of class of the instruction, data values involved in its evaluation and so on, as well as on its *execution context*, i.e., the set of instructions in $\mathcal{W}$ near to $\mathbf{w_n}$. The *execution context* of $\mathbf{w_n}$ can be divided in two contributions. The *first-order* contribution is due to the previous instruction $\mathbf{w_{n-1}}$, while the *second-order* contribution consists of the other instructions in the pipeline during the execution time of $\mathbf{w_n}$. In our model, we directly consider only the first-order contribution, while the second order one is only indirectly accounted for in terms of additional stall/latency cycles introduced during the execution of instruction $\mathbf{w_n}$.

We assume the design of the core as partitioned in a control unit and a data-path unit. The estimation of the energy consumed by the processor during the execution of the sequence $\mathcal{W}$ is given by:

$$E(\mathcal{W}) = \sum_{\forall n \in N} E(\mathbf{w_n}|\mathbf{w_{n-1}}) + E_c + c_0 \tag{4.3}$$

where the term $E(\mathbf{w_n}|\mathbf{w_{n-1}})$ is the energy dissipated by the data-path associated with the execution of instruction $\mathbf{w_n}$ dependent on the properties of $\mathbf{w_n}$,$\mathbf{w_{n-1}}$ as well as their *execution context*, E_c is the total energy dissipated by the control unit and the term c_0 is an energy constant associated with the start-up sequence of the processor (to reset the sequential elements of the processor).

In this work, one of our basic assumption is that the control unit is explicitly designed outside of the pipeline and it can be considered relatively invariant in terms of power consumption with respect to the type of instructions executed. Since, for the control units, accurate power estimates can be derived by using power models developed for finite state machines (such as those presented in [115]), we focus our research on the characterization at the instruction-level of the power consumption of the data-path.

Furthermore, we assume that the pipeline stages enjoy the *temporal additive property*, i.e. the energy consumption associated with an instruction flowing through the pipeline stages corresponds to the sum of the energy contributions of each pipeline stage. The property is based on the assumption that the processor function units associated with each pipeline stage are independent to each other.

Therefore, given the processor pipeline composed of a set S of stages, the energy consumption associated with $\mathbf{w_n}$ corresponds to the sum of the energy consumption of the modules that execute $\mathbf{w_n}$ in each one of the pipeline stages:

$$E(\mathbf{w_n}|\mathbf{w_{n-1}}) \approx \sum_{\forall s \in S} A_s(\mathbf{w_n}|\mathbf{w_{n-1}}) + I(\mathbf{w_n}|\mathbf{w_{n-1}}) \tag{4.4}$$

where:

- the term $A_s(\mathbf{w_n}|\mathbf{w_{n-1}})$ is the average energy consumed *per stage s* when executing instruction $\mathbf{w_n}$ after instruction $\mathbf{w_{n-1}}$;
- the term $I(\mathbf{w_n}|\mathbf{w_{n-1}})$ is the energy consumed by the connections between pipeline stages (inter-stage connections).

The function units used in more than one pipeline stage, such as the register file, can be considered only once per instruction with an average value that must be characterized separately.

To better detail the two terms A_s and I in Equation 4.4, we can consider that the execution of two consecutive instructions depends on two types of events. The first type of event is a data cache miss during the execution of an instruction: in this case the processor stalls all the instructions in the pipeline until the write or read data cache miss is resolved. Figure 4.4 shows an example of a data cache miss occurring during the execution of instruction E and another one during instruction G (the source of the stall is in the MEM stage). The second possible type of event is represented by an instruction cache miss during the fetch of an instruction: when this event occurs, the instruction cache generates to the core an 'IC-MISS' signal, that is used to freeze the inputs of pipeline registers, avoiding useless switching activity. Figure 4.4, shows an example of an instruction cache miss occurring during the fetch of instruction F and another one during the fetch of instruction G. In both cases, the cost for such misses consists of l extra cycles. The behavior of the pipeline is similar to the execution of a sequence of NOP operations, with the difference that, in case of NOPs, some energy is dissipated due to the flowing of the NOP within the pipeline registers. As a matter of fact, in the proposed model, the presence of a NOP instruction generated at compile-time, is considered as a normal instruction and it is characterized similarly to the other instructions in the ISA.

In our model, the average energy consumption *per stage* is given by:

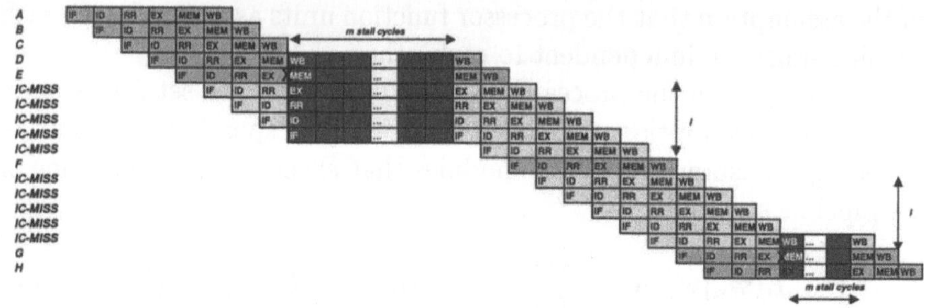

Figure 4.4. A trace of execution showing two data cache misses and two instruction cache misses

$$A_s(\mathbf{w_n}|\mathbf{w_{n-1}}) = U_s(\mathbf{w_n}|\mathbf{w_{n-1}}) + \sigma_s^n + \mu_s^n \tag{4.5}$$

where:

- U_s is the average energy consumption of stage s during an *ideal* execution of $\mathbf{w_n}$ in the absence of any hazards or exceptions, by assuming *one-cycle* execution per stage;
- σ_s^n is the additive average energy consumption by stage s due to a miss event on the data cache occurred when $\mathbf{w_n}$ is in s;
- μ_s^n is the additive average energy consumption by stage s due to a miss event on the instruction cache when it receives the IC-MISS signal.

The single terms appearing in Equation 4.5 are explained hereafter. Although Equation 4.5 still maintains a pipeline-aware feature, it can be characterized by a simple *linear regression* and it can be used also when there is not a precise view of the occurrence of cache misses, for example at the compiler level, where these statistics can be gathered only approximately.

The model for the energy consumption due to inter-stage connections I can be averaged as:

$$I(\mathbf{w_n}|\mathbf{w_{n-1}}) \approx \frac{1}{2} * V_{dd}{}^2 * f_{clk} * C_L * \alpha_{ave(\mathbf{w_n}|\mathbf{w_{n-1}})} \tag{4.6}$$

where V_{dd} is the power supply, f_{clk} is the pipeline clock frequency, C_L is the capacitive load of the inter-stage buses and $\alpha_{ave(\mathbf{w_n}|\mathbf{w_{n-1}})}$ is the switching activity of the inter-stage buses averaged with respect to the number of pipeline stages and the number of bit per stage N_s.

In the following, we detail the three terms appearing in Equation (4.5).

Class	Description
NOP	No operation
LS	Load Store data between registers and memory
ALU	Arithmetic-Logical instructions
MUL	Multiply instructions
CNT	Control-flow instructions

Table 4.1. VLIW operation classes considered for the target VLIW processor

3.1 Average Energy per Stage

The energy term U_s is influenced by the current instruction executed and by the previous one in the sequence. To detail the term U_s, we need to introduce the *spatial additive property* by assuming that the energy associated with each pipeline stage corresponds to the sum of the energy contributions due to the single operations in the current bundle, given the operations in the previous bundle. This property enables us to split the energy associated with a pipeline stage into the sum of the energy contributions due to each of the multiple and independent functional units that are active in the given stage. In our model, we apply this property not only to the EX stage, for which this concept appears intuitive, but also to all pipeline stages. An *energy base cost*, common to all executable bundles, has also been taken into account, as introduced in the pipeline energy model shown in the previous section.

Given the generic stage s, the additive property allows us to state that the term U_s can be expressed as:

$$U_s(\mathbf{w_n}|\mathbf{w_{n-1}}) = U_s(\mathbf{0}|\mathbf{0}) + \sum_{\forall k \in K} \nu_s(w_n^k|w_{n-1}^k) \qquad (4.7)$$

where the term $U_s(\mathbf{0}|\mathbf{0})$ is the base energy cost that represents the energy consumed by stage s during an execution of a bundle constituted entirely by NOPs ($\mathbf{0} = [\mathtt{NOP}\ldots\mathtt{NOP}]^T$), while the summation accounts for the terms $\nu_s(w_n^k|w_{n-1}^k)$, that are additional energy contributions due to the change of operation on the same lane k (w_n^k represents the operation issued on lane k by the long instruction $\mathbf{w_n}$).

Let us consider a 4-issue VLIW architecture with an ISA composed of five operation classes (see Table 4.1 for a summary of these classes), such as the architecture analyzed in Section IV. To estimate the cost U_s associated with an instruction $\mathbf{w_n}$ of the target architecture, first we need an estimate of the energy base cost $U_s(\mathbf{0}|\mathbf{0})$ for each stage of the pipeline, second we need an estimate of the energy contributions for the generic sequence of operations on the k-th lane $\nu_s(w_n^k|w_{n-1}^k)$ for each stage of the pipeline. Given the following sequence of two instructions:

$$\mathbf{w_n} = \begin{bmatrix} \texttt{ALU} \\ \texttt{NOP} \\ \texttt{NOP} \\ \texttt{NOP} \end{bmatrix}, \mathbf{w_{n-1}} = \begin{bmatrix} \texttt{LS} \\ \texttt{NOP} \\ \texttt{ALU} \\ \texttt{NOP} \end{bmatrix} \tag{4.8}$$

the overall cost associated with $\mathbf{w_n}$ for each stage would be:

$$U_s(\mathbf{w_n}|\mathbf{w_{n-1}}) = U_s(\mathbf{0}|\mathbf{0}) + \nu_s(\texttt{ALU}|\texttt{LS}) + \nu_s(\texttt{NOP}|\texttt{ALU}) \tag{4.9}$$

since, by definition, we know that $\nu_s(\texttt{NOP}|\texttt{NOP}) = 0$.

Generally, given an ISA composed of C operation classes, each ν_s is a look-up table composed of $\frac{C^2+C}{2}$ elements. In the case of standard instruction-level power characterization, we need a matrix composed of C^{2K} elements to characterize the inter-instruction effects between $\mathbf{w_n}$ and $\mathbf{w_{n-1}}$.

Let us now detail the terms σ_s^n and μ_s^n, considering the energy consumed by stage s whenever the number of cycles to execute $\mathbf{w_n}$ exceeds one since there is a miss either in the data or in the instruction cache.

3.2 Average Energy of a Data Cache Miss

The energy contribution σ_s^n considers the core energy consumption due to a data cache miss. In this case, the processor stalls all the instructions in the pipeline until the write or read data cache miss is resolved. This term depends on the probability per instruction that the processor stalls the pipeline:

$$\sigma_s^n = m_s^n * p_s^n * S_s \tag{4.10}$$

where m_s^n is the average number of additional cycles (stall cycles) occurred due to a data cache miss during the execution of the $\mathbf{w_n}$ in s, p_s^n is the probability that this event occurs, and S_s is the energy consumption per stage of the processor modules that are active due to a data cache miss. In particular, our model considers the energy cost of a data cache miss on the processor core, but we do not consider the energy cost of the cache and memory arrays when this event occurs.

3.3 Average Energy of an Instruction Cache Miss

Similarly, the energy contribution μ_s^n accounts for the additional energy spent by instruction $\mathbf{w_n}$ due to an instruction cache miss. When this event occurs, the instruction cache generates an IC-MISS signal as explained before. This contribution is given by:

$$\mu_s^n = l_s^n * q_s^n * M_s \tag{4.11}$$

where l_s^n is the average number of additional cycles (IC-MISS penalty cycles) occurred after the execution of the $\mathbf{w_n}$ in s due to an instruction cache miss, q_s^n is the probability that this event occurs, and M_s is the energy consumption per stage of the processor modules that are active due to an instruction cache miss. In particular, our model considers the energy contribution of an instruction cache miss on the processor core, while we do not consider the energy cost of the cache and memory arrays.

Let us consider an example of computation of $A_s(\mathbf{w_n}|\mathbf{w_{n-1}})$ when an Instruction Cache miss occurs after $\mathbf{w_n}$. The related effects must be taken into account because the IC-MISS signal is propagated within the pipeline and this affects the energy behavior of $\mathbf{w_n}$. If we refer to the example of Figure 4.4, this case corresponds to the evaluation of $A_s(E|D) = U_s(E|D) + l * q_s^E * M_s + m * p_s^E * S_s$ where $q_s^E = 1 \forall s$, and $p_s^E = 1$ for $s \in \{MEM, EX, RR, ID, IF\}$ and $p_{WB}^E = 0$.

3.4 Model Reduction for Scalar Pipelined Processors

The proposed energy model for VLIW pipelined processors can be simplified for *scalar* pipelined processors, where the spatial dimension is reduced to one. In this case, our model complies with the model of Tiwari *et al.* [99], that can be summarized as in Equation (3.22), where:

$$B_i = \sum_{s \in S} [U_s(0|0) + \nu_s(i|i)] \tag{4.12}$$

$$O_{i,j} = \sum_{s \in S} [\nu_s(i|j) - \nu_s(i|i)] \tag{4.13}$$

$$\sum_k E_k = \sum_{1 \leq n \leq N, s \in S} [\mu_s + \sigma_s] \tag{4.14}$$

3.5 Model Characterization

The proposed model depends on a set of energy parameters ($\mathbf{U}_s(\mathbf{0}|\mathbf{0})$, ν_s, M_s and S_s) that must be characterized once and for all by the supplier of the VLIW core.

Both $\mathbf{U}_s(\mathbf{0}|\mathbf{0})$ and ν_s can be characterized by re-writing the model represented by Equation (4.7) in the following matrix form:

$$\mathbf{U}_s = \mathbf{X}\beta_s \tag{4.15}$$

where $\mathbf{U}_s$ is a vector of E elements representing the observed energy consumption for E experiments, where each experiment e is a program composed of a sequence of the same pair of instructions $(\mathbf{w_{e,a}}, \mathbf{w_{e,b}})$.

Vector β_s is given by:

$$\beta_s = \begin{bmatrix} U_s(\mathbf{0}|\mathbf{0}) \\ \mathbf{\Theta}_s \end{bmatrix} \tag{4.16}$$

where $U_s(\mathbf{0}|\mathbf{0})$ is the scalar value representing the base cost of the instruction, while $\mathbf{\Theta}_s$ is a vector of T elements that linearly stores the elements of the upper diagonal part of the ν_s (since it is a symmetrical matrix) except for the term $\nu_s(\mathtt{NOP}|\mathtt{NOP})$. If C is the number of classes in the ISA, then $T = (\frac{C^2+C}{2}) - 1$.

Matrix $\mathbf{X}$ has a size of $E \times (T+1)$ and it is given by:

$$\mathbf{X} = \left[\begin{array}{c|c} 1 & \eta_1 \\ \cdots & \cdots \\ 1 & \eta_e \\ \cdots & \cdots \\ 1 & \eta_E \end{array}\right] \tag{4.17}$$

where η_e is a row vector of T elements associated with each experiment e such that:

$$\eta_e \times \mathbf{\Theta}_s = \sum_{\forall k \in K} \nu_s(w_{e,a}^k | w_{e,b}^k) \tag{4.18}$$

where "$\times$" represents the scalar product. Each element $\eta_e(i)$ stores the number of times that the i-th term of $\mathbf{\Theta}_s$ must be considered into the computation of Equation (4.18). This vector can be directly derived from the experimental setup chosen for experiment e.

The vector of parameters β_s can be estimated by means of a linear regression, where the rank of $\mathbf{X}$ must be of $(\frac{C^2+C}{2})$, that is, the design of the experiments requires at least $(\frac{C^2+C}{2})$ experiments whose pairs $(\mathbf{w_{e,a}}, \mathbf{w_{e,b}})$ are linearly independent. Given this lower-bound, for a K-issue VLIW processor with S pipeline stages and with an ISA composed of C classes of operations, the problem complexity of evaluating the energy parameters is reduced from $O(SC^{2K})$ to $O(SC^2)$, where C is the number of operation classes derived by clustering the ISA.

The other energy parameters (M_s and S_s) can be simply estimated by measuring the power consumption of the processor during, respectively, an I-cache miss and a pipeline stall.

3.6 Plug-in of the Model into an ISS

The application-dependent variables of the model must be estimated or computed during an instruction set simulation. These contributions can be gathered either in accurate mode or in statistically approximated mode.

In the first case, operations ordering and cache miss length can be dynamically computed by the simulator while the cache miss event is predicted either statistically ($p_s^n, q_s^n \in [0,1]$) or deterministically ($p_s^n, q_s^n \in \{0,1\}$). The accurate mode is used by an ISS elaborating the power model to compute on-the-fly the power consumption associated with the program.

In the statistically approximated mode, operations ordering, cache miss length and operation latencies are statistically averaged values used to generate an average off-line power profile of the given application. In this case, while p_s^n and q_s^n become averaged values $\overline{p_s}$ and $\overline{q_s}$ respectively, the term $\nu_s(w_n^k|w_{n-1}^k)$ is substituted by a weighted average value $\overline{\nu_s}$:

$$\overline{\nu_s} = \sum_{o_j^k, o_h^k \in ISA} p(o_j^k|o_h^k)\nu_s(o_j^k|o_h^k);$$

where $p(o_j^k|o_h^k)$ is the probability, estimated by the ISS, that operation o_j^k follows o_h^k in lane k.

4. Experimental Results

In this section, we describe how the energy model has been efficiently applied to characterize at the instruction-level the energy consumption of a VLIW-based customized embedded system. We developed an in-house VLIW-based system-level design environment (see Figure 4.5) composed of:

- A 4-issue VLIW processor with 6-stage pipeline characterized by an ISA composed of five operation classes (see Table 4.1).
- A configurable Instruction Cache.
- A configurable Data Cache.
- A 72 x 32-bit Register File.
- A main memory containing the object code to be executed by the VLIW processor.

The processor pipeline is composed of six stages (Instruction Fetch (IF), Instruction Decode (ID), Register Read (RR), Execute (EX), Memory access (MEM) and Write Back (WB)) and four lanes. Each lane can

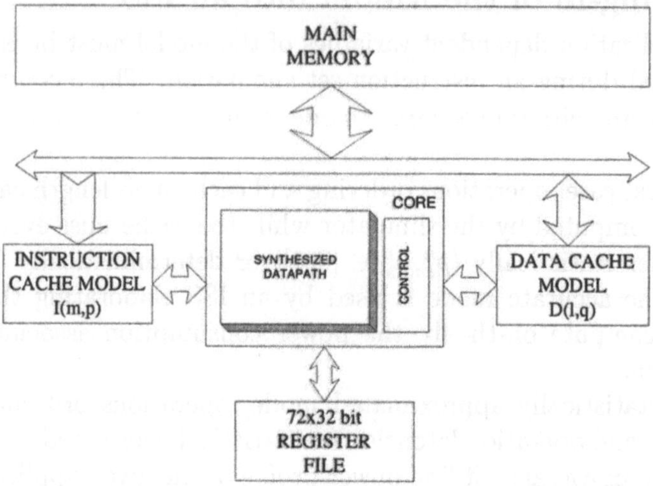

Figure 4.5. **The system-level test-bench used to validate the model.**

execute an integer operation (with a dedicated ALU), a multiply operation (with a dedicated multiplier) or a load/store operation trough a load/store unit common to all the lanes. The VLIW core supports data forwarding and operation latencies varying between one and two cycles (in the case of load/store operation) with in-order issue and in-order completion.

The configurable instruction cache $I(l,q)$ generates a miss signal for l cycles with an average probability q for each normal instruction fetched. The configurable data cache $D(m,p)$ can generate a miss with a probability p for each access and, during a miss, it stalls the VLIW pipeline for m cycles.

All design modules have been described by using Verilog-HDL. The data-path of the processor has been described at RT-level and it has been synthesized as multi-level logic by using Synopsys Design Compiler. The gate-level netlist has been mapped to a commercial 0.25μm and $2.5V$ technology library. Gate-level simulations of the core have been performed by using Synopsys VCS 5.2. We used PLI routines from Synopsys to retrieve toggle information for the core cells and we used the resulting toggle file as input to Synopsys Design Power tool. Energy estimates have been obtained by Synopsys Design Power tool at 100 MHz clock frequency.

4.1 Design of Experiments

The experimental design flow consists of two main phases: the first phase is the model characterization to derive the energy parameters, the second phase consists of the validation of the characterized model against the execution of embedded applications.

4.1.1 Experiments for Model Characterization

Based on Equation (4.15), an automatic tool has been defined to generate a set of 250 experiments, where each experiment e is composed of a sequence of 1000 pairs of instructions $(\mathbf{w_{e,a}}, \mathbf{w_{e,b}})$. The automatic tool satisfies the requirements imposed on the rank of the matrix of experiments $\mathbf{X}$ (see Equation (4.15)). As a matter of fact, given $C = 5$ we can easily reach $rank[\mathbf{X}] = 15$.

The experiments have been generated by varying the following parameters:

- Number and type of operations in the pair $(\mathbf{w_{e,a}}, \mathbf{w_{e,b}})$.
- Registers used in the instructions (registers are addressed through 7-bit addresses).
- Values of the 9-bit immediate operands used in the instructions.

Finally, the energy consumption of the stages of the processor during stalls and I-cache misses have been measured by directly gathering toggle information.

4.1.2 Experiments for Model Validation

The model validation phase consists of four main steps.

- Step 1 represents the model validation when neither I-cache misses nor D-cache misses occur. With this purpose, we generated a set of 250 experiments, each one composed of a sequence of 1000 pairs of instructions $(\mathbf{w_{e,a}}, \mathbf{w_{e,b}})$.
- Step 2 represents the model validation during pipeline stalls. A set of experiments has been generated such that the average stall probability per instruction (forced by $D(m,p)$) ranges from 1% to 14% (according to [51], this is the typical miss rate range for a generic cache). The parameter m, in this case, has been set to 25 cycles.
- Step 3 consists of the model validation during I-cache misses. In this case, another set of experiments has been generated such that the average miss rate of $I(l,q)$ ranges from 1% to 14%. The parameter l, in this case, has been fixed to 36 cycles.

$\sum_{s\in S}\nu_s$	NOP	AL	MUL	DT	CNT
NOP	0	339	350	349	172
AL	-	270	363	450	385
MUL	-	-	429	424	467
DT	-	-	-	411	467
CNT	-	-	-	-	80

Table 4.2. Characterization results for the matrix ν summed over the pipeline stages for all pairs of operations when neither pipeline stalls nor I-cache misses occur. Energy values are given in [pJ].

Pipeline Stage	Avg.Err.	Std. Dev.	Max. Err.
IF	4,24%	3,15%	13,69%
ID	1,27%	1,13%	8,06%
RR	3,54%	2,02%	11,17%
EX	6,75%	4,16%	28,26%
MEM	5,43%	5,81%	26,19%
Interconnect	13,59%	14,33%	116,91%

Table 4.3. Comparison results of measured vs. estimated power values for each pipeline stage and for the interconnection.

- Step 4 consists of the model validation against a set of kernels derived from embedded DSP applications. The kernels are characterized by several behaviors in terms of instructions composition and I- and D-cache misses.

4.2 Model Characterization

First, the proposed core model has been characterized to derive the energy parameters. The measured energy base cost $\mathbf{U}_s(\mathbf{0}|\mathbf{0})$ summed over all the pipeline stages is 1490.57 [pJ]. Table 4.2 reports the characterization results for the matrix ν summed over the pipeline stages for all pairs of operations when neither pipeline stalls nor I-cache misses occur. The value of M_s and S_s parameters summed over the pipeline stages are 1400 [pJ] and 110 [pJ] respectively.

4.3 Model Validation

The experimental results derived during Step 1 have been reported in Table 4.3 and Fig. 4.6. Table 4.3 shows the comparison results of measured power values with respect to estimated power values for each pipeline stage and for the interconnection network. Since the RF power consumption has not been considered (being the RF outside of the core) and the pipeline latch MEM/WB is contained in the module of pipeline stage MEM, the power consumption due to the WB stage is considered as zero and therefore it has been omitted. Fig. 4.6 reports the agreement between the measured and the estimated power values for the in-house VLIW core, when neither I-cache misses nor D-cache misses occur. For the given set of experiments, the absolute maximum error is 11.4%, the absolute average error is 3.19%, and the standard deviation is 2.07%.

Concerning Step 2, Fig. 4.7 shows the comparison between measured and estimated power values when pipeline stalls occur. Globally, the absolute maximum error is 20.2% and the absolute average error is 3.1%. The standard deviation of the error is 2.7%. The lowest power values are associated with experiments with a higher data cache miss probability: in fact, as the miss probability increases, the power consumption reaches the stall power consumption S_s summed over the pipeline stages.

Regarding Step 3, Fig. 4.8 reports the agreement between measured and estimated power values when I-cache misses occur. The absolute maximum error and the absolute average error are 9.6% and 3.2% respectively. The standard deviation of the error is 2.2%. In this case, the lowest values of power consumption are associated with experiments with a higher instruction cache miss probability. In fact, as the miss probability grows, the power consumption approaches the instruction cache miss power consumption M_s summed over the pipeline stages.

Finally, for Step 4, Fig. 4.9 compares the measured and estimated power values for the given set of DSP benchmarks, where both pipeline stalls and I-cache misses parameters can vary. The given kernels (written is assembly code for the target processor) are:

- An unoptimized DCT and IDCT transform applied to a 64 half-word block;
- An optimized version of the DCT;
- An optimized version of the IDCT;
- A FIR routine (32-tap, 16-bit);
- A optimized (unrolled) FIR routine (32-tap, 16-bit);
- A gaussian elimination algorithm;

The maximum absolute power error is 10%, while the average absolute power error is 4.8%. Full core estimation at the gate-level is four orders of magnitude slower than instruction-level simulation. In our set of experiments, an under-estimation effect has been shown, that can be due to the approximation used in our characterization process to capture the data behavior of single instructions.

As a final example, we considered the error estimates generated when both D-cache and I-cache misses are not considered in the model. In this case, the energy for each instruction A_s becomes the ideal energy U_s, so we would expect an under-estimation effect on the power consumption. However, the total energy computed is divided by the ideal execution time, that is much less than the real execution time giving an over-estimation of the actual power value.

Figure 4.10 reports the percentage errors between the predicted power values and the measured power values when stall events are not considered. Miss probability values are the same used for the characterization and validation of the model. If stall events are not considered in the model, the predicted length of one instruction is ideal, and the predicted instruction energy consumption (even not considering the energy lost during stalls) is related to only one cycle, providing an overestimation effect on the power consumption that is almost proportional to the stalls per instruction rate.

Finally, Figure 4.11 shows the percentage errors existing between the estimated power values and the measured ones, when instruction cache miss events are not considered. Miss probability values are the same used for the characterization and validation phases of the model. If I-cache miss events are not considered in the model, the predicted length of one instruction is ideal, and the predicted instruction energy consumption is related to only one cycle resulting, even in this case, in a power over-estimation that is almost proportional to I-cache misses per instruction rate.

5. Conclusions

In this chapter, we proposed an instruction-level energy model for VLIW pipelined architectures. The model has been applied to a 4-issue VLIW pipelined custom processor during the execution of a given set of DSP-based multimedia benchmarks. The proposed model affords the problem of the estimation of the power budget for VLIW-based single-cluster architectures, where all operations in a long instruction access the same set of functional units and the same register file. On-going works aim at exploring the power budget for multi-clustered VLIW architectures. Other future directions of the work aim at integrat-

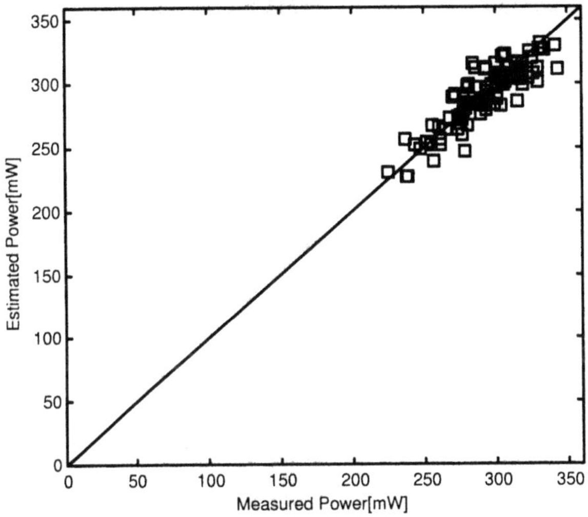

Figure 4.6. Agreement between measured and estimated power values when neither I-cache misses nor D-cache misses occur. (Absolute maximum error 11.4%).

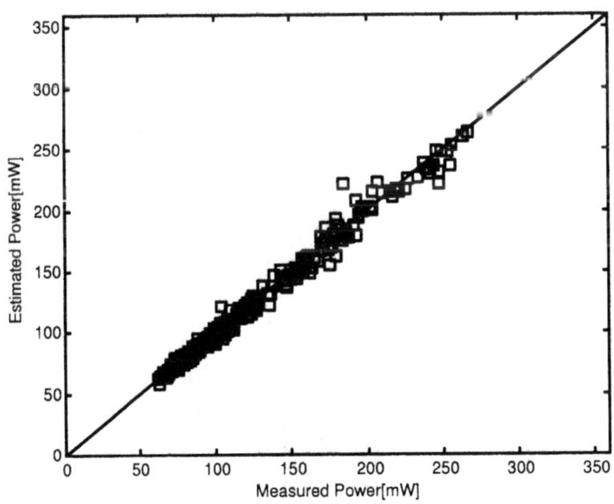

Figure 4.7. Agreement between measured and estimated power values when D-cache misses occur. (Absolute maximum error 20.2%).

ing the proposed energy model in a more general system-level exploration methodology to evaluate the impact of architectural parameters on energy-performance standpoints. Furthermore, our instruction-level

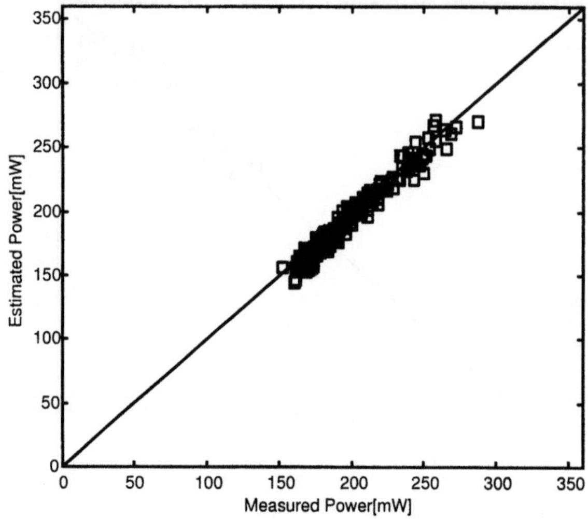

Figure 4.8. Agreement between measured and estimated power values when I-cache misses occur. (Absolute maximum error 9.6%).

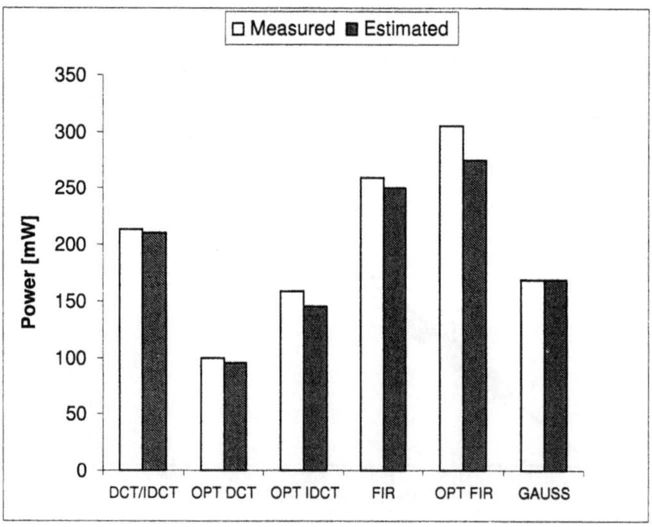

Figure 4.9. Comparison between measured and estimated power values for the given set of DSP benchmarks. (Absolute maximum error 10%).

energy model can be used to define microarchitectural optimizations of the processor as well as to assess compilation optimizations such as instruction scheduling and register labeling.

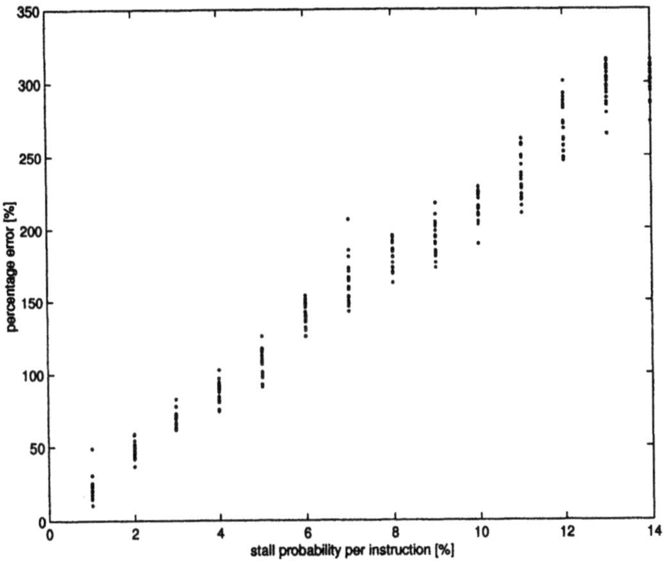

Figure 4.10. Errors generated not considering D-cache miss events

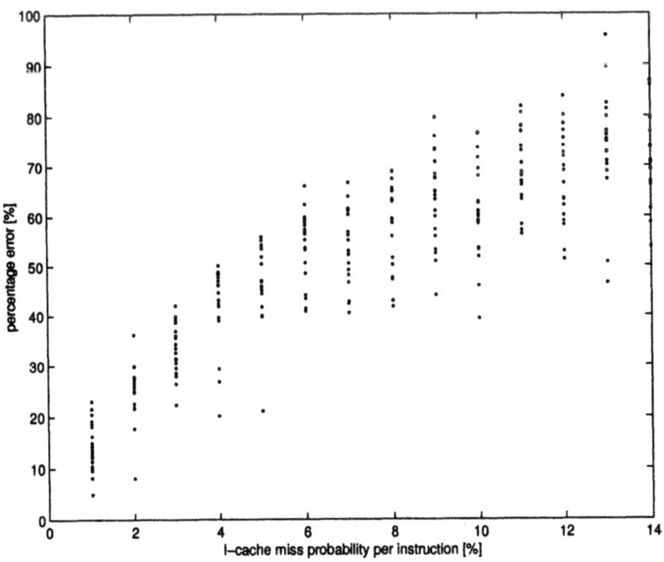

Figure 4.11. Errors generated not considering I-cache miss events

Chapter 5

SOFTWARE POWER ESTIMATION OF THE LX CORE: A CASE STUDY

In this chapter we present an application of the introduced energy model, to theLx processor, a commercial 4-issue VLIW core jointly designed by HPLabs and STMicroelectronics. As we said previously, the characterization strategy takes into account several software-level parameters and provides an efficient instruction-level power model based on instruction clustering techniques. The accuracy of the proposed model has been qualified against a industrial gate-level simulation-based power estimation engine. The experimental results show an average error of 1.9% between the instruction-level model and the gate-level model, with a standard deviation of 5.8%. We conclude the chapter by showing how the presented power estimation methodology has been succesfully applied to explore the power consumption at the software level by constructing a brand new horizontal execution-scheduling algorithm providing an average energy saving of 12%.

Furthermore, we show how the proposed model has been extended to provide early power figures and energy/performance trade-offs of a future multi-clusterd architecture of the same target processor.

1. Introduction

The *first* goal of this chapter is to present the application of the power model, proposed in chapter 4, to a real industrial processor: the Lx architecture, a scalable and customizable processor technology [116] designed for multimedia and signal processing embedded applications.

The *second* goal is to show how the complexity of the energy model can be further simplified by introducing a methodology to automatically cluster the whole Instruction Set with respect to their average energy cost, to converge to an effective design of experiments for the actual

V. Zaccaria et al. (eds.),
Power Estimation and Optimization Methodologies for VLIW - based Embedded Systems, 67–85.

characterization task. The *third* goal is to apply the instruction-level power model developed so far, not only to provide data on the power consumption of the core, but also to statically optimize for power the embedded software applications in order to obtain power savings. The proposed technique consists of an *horizontal scheduling* algorithm similar to the one presented in [117]. The basic idea consists of rescheduling the parallel operations within the same bundle based on the power figures carried out from the power characterization phase, without increasing the overall latency associated with the code. The main difference of the approach proposed here with respect to the approach proposed in [117] is that, in our case, the effectiveness of a specific scheduling policy is quantified based on the actual power model, while in [117] the goodness of a pair of scheduled operations is evaluated in terms of the Hamming distance between the coding of two adjacent operations carried out in the same processor's lane. The *fourth* and final goal, is to provide early power figures and energy/performance trade-offs of a future multi-clusterd architecture of the same target processor. This has been done by realizing two different kind of analyses: a *non constrained analysis*, based on the *energy-delay product*, and a *constrained analysis* that reports the energy/delay pareto curves of the various multi-clustered system configurations. These results should help the designer to asses the best configuration that optimizes both the performances and the energy consumption in the early stages of the multi-cluster design.

The chapter is organized as follows. An overall description of the customization applied to the proposed VLIW power model is given in Section 2, while Section 3 introduces the automatic methodology to cluster operations with the same power behavior. The proposed horizontal scheduling algorithm aims at saving power, while maintaining performance is described in Section 4. Some experimental results demonstrating the applicability of the proposed general estimation and optimization approach to the industrial Lx VLIW processor are reported in Section 5. Finally, Section 6 contains some concluding remarks and outlines some future research directions originated from this work.

2. The VLIW Power Model

The instruction-level *(IL)* power model proposed in chapter 4 is accurately determined by taking into account all the constituent microarchitectural features of the processor, considering the effects that the individual very long instruction (namely bundle) can produce on them. This level of detail cannot be achieved by using a simple black-box instruction-level energy model such as those presented in the literature so far. In fact, the proposed model decomposes the energy contributions of a single macro-block in the energy contributions of the functional units that work separately on each operation of the bundle. This property, introduced as the *spatial additive property*, has been fundamental to deal with the complexity of the *IL* power model for VLIW cores, which grows exponentially with the number of possible operations in the *ISA*. The proposed decomposition provides a way to create a mapping between bundles and micro-architectural functional units involved during the simulation. This instruction-to-unit mapping is used by the ISS to retrieve the energy information *on the fly* to eventually compute run-time power estimates.

The power model proposed in Section 3 assumes that the energy associated with $\mathbf{w_n}$ is dependent on its own properties (e.g., class of the instruction, addressing mode and so on) as well as on its *execution context*, i.e., the former instruction $\mathbf{w_{n-1}}$ and the additional stall/latency cycles introduced during the execution of the instruction.

In particular, the power model can be summarized as follows:

$$E(\mathcal{W}) \approx \sum_{1 \leq n \leq N} \sum_{\forall s \in S} \Big[U_s(\mathbf{0}|\mathbf{0}) + \sum_{\forall k \in K} \nu_s(w_n^k | w_{n-1}^k) + \\ + m_s^n * p_s^n * S_s + l_s^n * q_s^n * M_s \Big] \quad (5.1)$$

Where the term $U_s(\mathbf{0}|\mathbf{0})$ is the base energy cost that represents the energy consumed by stage s during an execution of a bundle constituted entirely by NOPs ($\mathbf{0} = [\mathtt{NOP} \ldots \mathtt{NOP}]^{\mathrm{T}}$), $\nu_s(w_n^k|w_{n-1}^k)$ is the additional energy contribution due to the change of operation on the same lane k, m_s^n (l_s^n) is the average number of additional cycles due to a data (instruction) cache miss during the execution of the $\mathbf{w_n}$ in s, p_s^n (q_s^n) is the probability that this event occurs, and S_s (M_s) is the energy consumption per stage of the processor modules that are active due to a data (instruction) cache miss.

The model expressed by equation (5.1) can be customized, while preserving its accuracy, by means of some basic observations. The target VLIW processor is a pipelined processor, in general it could be quite difficult to isolate the power contributions due to the different processor's modules and pipeline stages. This aspect could represent a limit for the application of equation (5.1) but, as a matter of fact, the model can be tailored to further support these cases. The term $\sum_s U_s(\mathbf{0}|\mathbf{0})$ corresponds to the power consumption of the core while it is executing NOPs and can be substituted by the average base cost $U(\mathbf{0}|\mathbf{0})$. Besides, $\sum_s \sum_k \nu_s(w_n^k|w_{n-1}^k)$ can be substituted by a cost dependent only on the pair of instructions ($\sum_k \nu(w_n^k|w_{n-1}^k)$), that corresponds to the energy consumption of the core while it is executing the same pair of instructions $(\mathbf{w_n}, \mathbf{w_{n-1}})$. For instruction and data cache misses, we assume that after a transient state, the probabilities per stage (p and q) and their penalties (m and l) can be averaged for each instruction of the stream:

$$\sum_{s}(m_s^n * p_s^n * S_s + l_s^n * q_s^n * M_s) \rightarrow (m * p * S + l * q * M) \qquad (5.2)$$

where $m(l)$ is the average data (instruction) cache miss length, $p(q)$ is the average probability per stage and per instruction that a data (instruction) cache miss can affect one instruction; S (M) is the average energy consumption of the *entire* processor during these events. We will show in the experimental results that this assumption does not involve a significant loss in the accuracy.

The equation is thus reduced to:

$$E(\mathcal{W}) \approx \sum_{1 \leq n \leq N} \Big[U(\mathbf{0}|\mathbf{0}) + \sum_{\forall k \in K} \nu(w_n^k | w_{n-1}^k) + \\ + m * p * S + l * q * M \Big] \qquad (5.3)$$

The complexity of the model remains quadratic with respect to the number of operations within the instruction set $(O(K * |ISA|^2))$, while a black-box model would have a complexity of $O(|ISA|^K)$. This leads to a large reduction of the characterization effort, since the number of experiments to be done is reduced exponentially.

In the next section, we show how clustering the operations of the ISA with respect to their average energy consumption enables a faster and more effective characterization of the core's power consumption, while preserving the model accuracy.

3. Clustering of the Operations

In this work, we consider two operations in the *ISA* as "not equal" if they differ either in terms of functionality (i.e., opcode) or in terms of addressing mode (immediate, register, indirect, etc.). Even without considering data differences (in terms of register names or immediate values), the number of operation's pair to be considered in equation (5.3) would be too large to be characterized with a transistor-level or even gate-level simulation engine. Considering the Lx processor, for example, the *ISA* is composed of 70 operations, but the possible addressing modes for each operation would imply a characterization of 6126 pairs of operations to completely define the model. In particular, this is due to the characterization of the matrix ν (see equation (5.3)), while the other parameters do not depend on the particular instruction executed.

To give a rough idea of the time required for the characterization, a Sun Ultra Sparc 60 at 450MHz with 1GB RAM performs a gate-level power simulation of the Lx core within 25 minutes. Given such simulation time, we would need approximately 108 days to perform the complete characterization phase!

To sensibly reduce the number of experiments to be generated during the characterization phase, in this chapter we introduce the cluster analysis on the operations of the *ISA*. The basic idea of the cluster analysis consists of grouping in the same cluster the operations showing close energy cost to each other. The energy cost of an operation is defined as the energy consumed by the processor when it executes only that operation.

Various clustering algorithms appeared in literature so far (see [118, 119] for a survey); among these we have chosen the *k-mean clustering algorithm* to cluster energy values, since it requires a lower computational cost with respect to other methods such as hierarchical clustering.

Given a population $\Theta = \{e_1 \dots e_t \dots e_\theta\}$, where each e_t is the energy consumption of a single operation t, the k-mean clustering algorithm tries to partition Θ into a set of K clusters $(C_1 \dots C_K)$ to minimize the following mean-square error:

$$\sum_{j=1}^{K} \sum_{i=1}^{n_j} (x_{i,j} - c_j)^2 \tag{5.4}$$

where n_j is the number of elements of cluster C_j, $x_{i,j}$ is the i-th element of cluster j, c_j is the center of gravity of the j-th cluster and $\bigcup_j C_j = \Theta$.

Provided the number K of classes in which the original population must be partitioned, the k-mean clustering algorithm attempts to randomly split the whole set into K subsets. Then, each element is moved into the subset j whose center of gravity is closest. This procedure iterates until the stopping criterion is met.

We expect that operations using the same functional unit of the processor are characterized by a very similar power cost and, thus, they fall in the same cluster. As an anticipatory preview of the operations clustering on the Lx processor, Figure 5.1 shows four operations, two of which use the adder (`add, sub`), while the other two use the multiplier (`mull, mullhu`). For each operation, the different columns in Figure 5.1 correspond to different experiments created by varying randomly register names and values. As can be seen, the k-mean algorithm generates two clusters, each of which contains the operations that use the same functional unit.

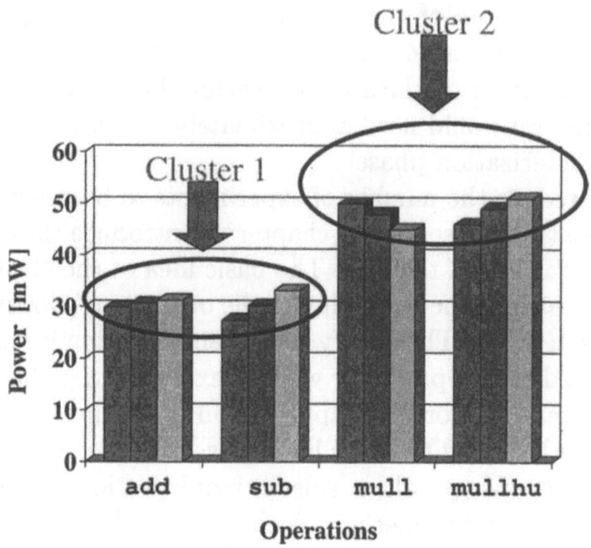

Figure 5.1. Similarity in the power cost of some basic Lx operations.

Once the instructions have been clustered into a set of $C_1 \ldots C_j$ clusters, we compute the matrix ν of equation 5.3 in the following way:

$$\nu(w_n^k | w_{n-1}^k) = \begin{cases} E_i & \text{if } w_n^k = w_{n-1}^k \wedge w_n^k, w_{n-1}^k \in C_i \\ D_{i,j} & \text{if } w_n^k \neq w_{n-1}^k \wedge w_n^k \in C_i, w_{n-1}^k \in C_j \end{cases} \tag{5.5}$$

Without considering the switching activity due to data dependency, this decomposition tries to model the fact that when operations are equal ($w_n^k = w_{n-1}^k$), they generate the least switching activity possible (first case). In the second case, when the operations are different, they generate an increased switching activity even if they are in the same cluster (accounted for by the matrix $D_{i,j}$). Note that the complexity of matrix ν has been reduced from $O(|ISA^2|)$ to $O(|C^2|)$.

To apply the clustering algorithm, one of the most relevant parameters to choose is the number of clusters k. Typically, this is a user-defined parameter and no automatic rules exist for that. As a general rule, the number of clusters is determined by a tradeoff between the maximum standard deviation of the elements within the same cluster and the number of experiments that must be done in order to characterize the model.

The minimum number of experiments is a typical cost function that must be reduced when performing the design of the experiments. In this work, we assume that the linear regression [84] is used to perform the characterization of the model. In this case, the minimum number of experiments is linearly dependent on the number of parameters of the model. Thus, the characterization cost has the following upper bound:

$$O\left(\frac{K \cdot (K-1)}{2}\right) \quad (5.6)$$

where K is the number of clusters.

Section 5 reports the application of the tradeoff analysis performed on the Lx processor to determine the number of clusters.

4. Horizontal Scheduling for Low-Power

Once the model has been fully characterized, the basic idea consists of using it as a cost function to statically re-order instructions to minimize the power associated with the executable application. Our work tries to overcome the limitations of the work proposed by Lee *et al.* [117], targeting the reordering of instructions based on the minimization of the switching activity in the instruction bus. This is an extremely exotic assumption indeed, since the switching activity of the instruction bus (due to the encoding of the register names or the operations) could be, in many cases, misleading in the determination of the overall switching activity (and thus power consumption) of the core. For example, two adjacent operation codes (i.e., with an Hamming distance equal to one) can be simply unrelated in terms of power consumption.

Our rescheduling algorithm considers each basic block of the code generated by the compiler and tries to reschedule operations within the same bundle (*horizontally*), without significantly increasing the overall latency associated with the code. This aspect is very important, since the goal is to obtain low-energy code execution without sacrificing performance.

Ideally the algorithm starts from bundle $\mathbf{w_1}$ and tries to find a suitable reordering of each w_n^k in order to minimize the characterized cost function $\sum_k \nu(w_n^k|w_{n-1}^k)$, being w_{n-1}^k fixed (see equation 5.3). This step is sequentially repeated on all the consecutive bundles $\mathbf{w_n}$ of the basic block. All the possible permutations of operations within the bundle are fully checked and no heuristics are applied.

The algorithm for low power scheduling is represented in Figure 5.2. The algorithm receives as inputs a list of bundles to be re-scheduled (*unscheduled*) and the cost function $\nu(w_n^k|w_{n-1}^k)$ and generates a list of re-scheduled bundles (*scheduled*). The basic operation of the algorithm

```
inputs:   unscheduled[]          array of bundles before low power scheduling
          v(w[k,n],w[k,n-1])     cost matrix for consecutive bundles
outputs:  scheduled[]            array of low power scheduled bundles

index = 0
old_bundle = unscheduled[index]
old_bundle = insert_nop(old_bundle)
scheduled_[index]=old_bundle
while(index<no_bundles)
{
     index++;
     current_bundle  = unscheduled[index]
     current_bundle  = insert_nop(current_bundle)
     bipartite_graph = create_bipartite_graph(old_bundle, current_bundle)
     matching = find_best_bipartite_matching(bipartite_graph,v)
     current_bundle = reorder(old_bundle, current_bundle, matching)
     scheduled[index]=current_bundle
     old_bundle = current_bundle
}
```

Figure 5.2. The algorithm for horizontal low power scheduling

is the construction of a bipartite graph (shown in Figure 5.3) that is composed of a set of edges connecting all the operations of the previous bundle (to which a lane has been already assigned) and all the operations of the current bundle. Each edge connecting two operations u and v is provided with a weight $\nu(u|v)$, as if the two operation u and v were put on the same lane. The algorithm then chooses a set (or *matching*) of K edges (K=number of lanes) connecting different nodes and subjected to hardware constraints whose sum of costs is minimum (*find_best_bipartite_matching*). The selection is performed by checking for all the possible permutations of the operations of the current bundle. Depending on the matching, the operations within the bundle are then reordered (*reorder* function) to produce the new bundle that is inserted in the final schedule. The function *insert_nop()* is used to fill bundles composed less that K operation with NOP operations in order to equalize all the bundles to a size equal to K operations. This is done to avoid that the Instruction Issue Unit of the Lx processor changes the lane assigned to each operations by the compiler.

The fact that each permuatation is checked by the scheduler does not inhibit the practical implementation of the algorithm on real VLIW cores (characterized in general by a number K of parallel operations in the range from 2 to 8) since the total number of permutations to be checked is $K!$. For example, for the 4-issue VLIW Lx processor core, the total number of permutations to be checked is limited to $4!=24$.

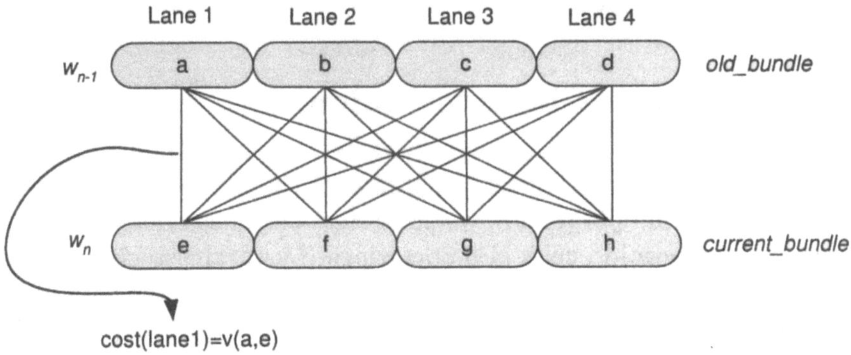

Figure 5.3. The bipartite graph generated from the current and the preceding bundle

5. The Lx Case Study

The Lx architecture is a scalable and customizable processor technology [116] jointly designed for multimedia and signal processing embedded applications. The Lx processor is a statically scheduled VLIW architecture designed by Hewlett-Packard and STMicroelectronics, that supports a multi-cluster organization based on a single PC and a unified I-cache. The basic processor is 4-issue VLIW core featuring four 32-bit integer ALUs, two 16x32 multipliers, one load/store unit and one branch unit. The cluster also includes 64 32-bit GPRs an 8 1-bit Branch Registers. Lx has an in-order 6-stage pipeline and a very simple integer RISC *ISA*. For the first generation, the scalable Lx architecture is planned to span from one to four clusters (i.e., from 4 to 16 instructions issued per cycle).

Lx is provided with a complete software tool-chain, where no visible changes are exposed to the programmer when the core is scaled and customized. The tool-chain includes a sophisticated *ILP* compiler technology (derived from the Multiflow compiler [120]), GNU tools and libraries. The Multiflow compiler includes both traditional high-level optimization algorithms and aggressive code motion technology based on trace scheduling.

A mix of RTL and gate-level netlist of the core processor has been used to perform the characterization needed for the power model presented in this work. The experiments have been carried out by using Synopsys VCS 5.2 and a set of PLI routines to elaborate toggle statistics over the whole gate-level netlist. PowerCompiler (by Synopsys) has been used to combine the toggle statistics with the power models of the standard cells library provided by STMicrolectronic, to compute the power figures of the entire core.

5.1 Lx Power Characterization

As a first step for the Lx power characterization, we first proceed to group the operations in clusters by means of the k-mean clustering algorithm. For each operation o, we generate a set of assembly programs composed of repeated cycles of o operations (by varying register names and values) and we measured the energy consumption of the core at the gate-level. Then we applied the k-means clustering algorithm for several values of K. In order to determine the most suitable number of clusters, we analyzed the minimum number of experiments that would be needed to characterize the model with K clusters and we performed a tradeoff analysis with respect to the maximum variance within the clusters. As mentioned above, the minimum number of necessary experiments depends linearly on the number and the size of the parameters involved in the model. In this case, the shape of the curve defining the number of experiments is quadratic, due to the quadratic dependence of the size of the matrix ν with respect to the number of clusters K.

In general, a small number of clusters implies a high variance within them (i.e., poor accuracy of the model), though it also implies a small number of experiments during the regression. On the contrary, a large number of clusters implies good accuracy, but would result in a huge number of experiments.

Figure 5.4 shows the results concerning the maximum variance within the operations clusters and the number of experiments required to characterize each corresponding model. We can note how, for a number of clusters equal to 11, we can obtain a good tradeoff between the maximum variation that drops to 13% and the minimum number of experiments (that reaches 78). For this reason, we selected 11 clusters to characterize the Lx's power model.

The value found for the experiments is a minimum value because, for each cluster, we can perform multiple experiments by varying the operations within the cluster, the register names and the immediate values.

Once selected the number of clusters and, therefore, the number of coefficients in the model (i.e., the size of the ν matrix), we realized a set of experiments in which each possible pairs of clusters could be generated more than once, by varying register names and values.

The experiments are characterized by a value of p and q (the data and instruction cache miss probabilities per instruction) as small as possible to accurately characterize by regression only $U(\mathbf{0}|\mathbf{0})$ and $\nu(w_n^k|w_{n-1}^k)$.

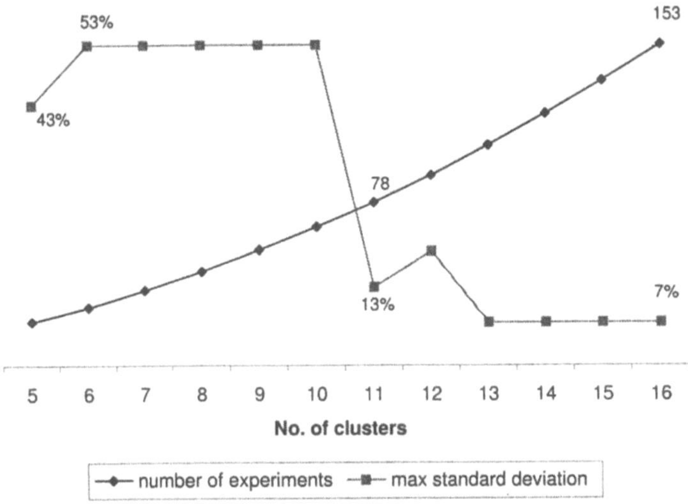

Figure 5.4. Tradeoff between the number of experiments required to characterize the model and the maximum variance within each cluster.

The values of M and S (the energy consumed in a cycle during an I-cache and D-cache miss) have been characterized by generating a set of few experiments with a large number of data and instruction cache misses, and by measuring the power only during these events. Finally, the m and l miss penalties have been extrapolated by looking at the behavior of the microarchitecture during these events.

The model has been validated with a set benchmarks, different from the experimental setup used to perform the characterization. The validation benchmarks are composed by a sub-set of the Mediabench applications [121] (namely, g721 encoder and decoder, epic encoder and decoder, mpeg2), a set of finite impulse response filters, discrete cosine transforms and matrix elaboration algorithms. Figure 5.5 shows the agreement between the gate-level power values and the power values estimated with the proposed model. Here, the parameters for the instruction level model, i.e., probability per instruction per lane as well as data and instruction cache miss probabilities, have been gathered by RT-level simulation so they are the *actual* (or *ideal*) values. In these conditions the power model has shown, on the validation benchmarks, a mean error of 1.9% and a standard deviation on the error of 5.8%. The multiple correlation coefficient, that explains the percentage of the total variation explained by the model, has been computed as in [84] and is equal to:

$$\sqrt{\frac{SSR}{SSR+SSE}} = 90\% \tag{5.7}$$

As a matter of fact, there are only three benchmarks whose error on the prediction is in the neighborhood of 10% probably because of the high switching activity of the data consumed by instructions that is not captured by the model. For high-level/instruction-level power macro models this can be considered an acceptable value in terms of accuracy. However, the plug-in of the model into an ISS can introduce some error due to the inaccuracy of the ISS to model the exact behavior of the lanes and cache misses. This issue is analyzed in the next chapter.

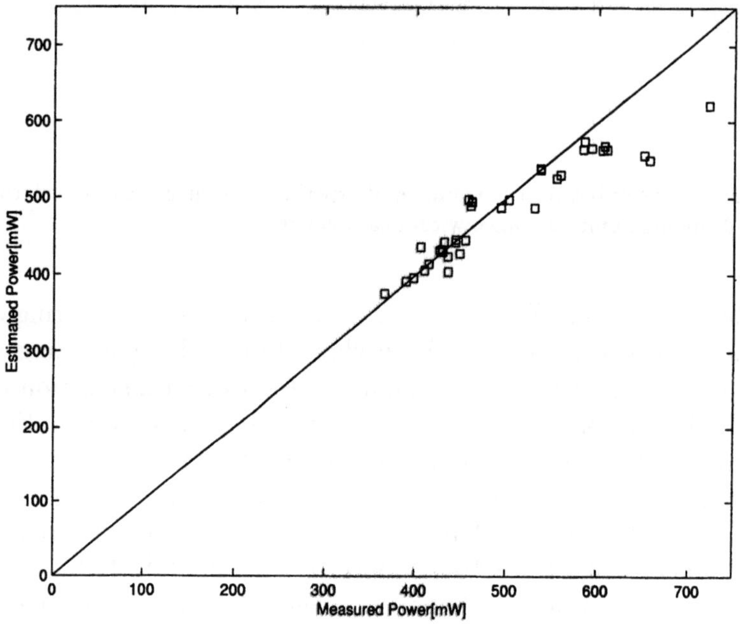

Figure 5.5. Scatter plot of the measured power values with respect to the estimated power values

5.2 Lx Power Optimization

The Lx power model has then been used to find out a good schedule for low-power the Lx assembly code of two FIR filters (*fir1* and *fir2*), the fast discrete cosine transform (*fastdct*), the matrix multiply algorithm (*matrix*) and the bubble sort algorithm (*sort*). The rescheduling algorithm works as explained in section 4, by receiving as input the assembly code of the original benchmark generated by the Lx compiler. Figure 5.6 shows the percentage of power and energy savings obtained by applying the rescheduling algorithm to the selected set of benchmarks (positive percentages indicate an effective decrease in power/energy consumption). The power savings have been computed by simulating at the gate-level the execution of both the original and the *re-scheduled* version of the code for each test-case.

The measured average power consumption decreases by 17%, while the average energy consumption decreases only of 12%, since the rescheduling algorithm causes a slight increase in the latency of the code. This is due to the fact that operation rescheduling could impact the efficiency of the Lx's instruction compression mechanism, leading to an increment of cache misses. However, in some cases (*fir2* and *matrix*) the instruction cache misses are reduced since, probably due to the particular structure of the code, the rescheduling algorithm can lead to a more regular code in terms of instruction cache access patterns.

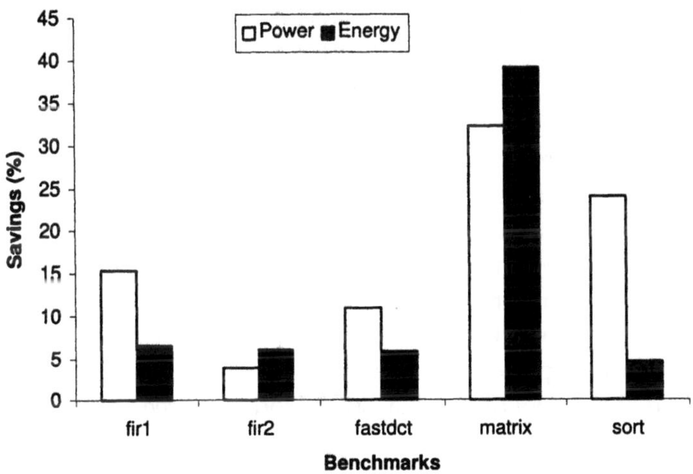

Figure 5.6. Power and energy savings obtained by our horizontal rescheduling algorithm applied to the Lx processor executing the selected set of benchmarks.

5.3 Early Design Space Exploration

The Lx is a customizable processor architecture that has been designed in order to be extended in a multi-clustered fashion. A multi-clustered Lx implementation has a *single instruction cache*, an *instruction fetch issue unit* and a set of cores (clusters) that are identical, in terms of function units, to the mono-cluster core. The role of the *in-*

struction fetch mechanism is to fetch a multi-cluster bundle from the instruction cache and dispatch the syllables to the appropriate clusters. Clusters have a separate register file and communicate through an *inter-cluster communication mechanism* that supports send and receive operations. Concerning the *data cache*, the specification for the multi-clustered architecture allows several possible configurations, among wich we can find a single multi-port cache shared between all clusters. We will assume this type of configuration for the remaining part of the chapter.

In this section we show how the instruction-level power model of Eq. 5.3 can be extended to forecast the power consumption of such multi-clustered processor. This analysis can be performed without having a detailed description (e.g. an HDL netlist) of the processor, thus it can be performed in the early stages of the design of an embedded system. This is of extreme help to the designer who can assess the suitability both in terms of power and performance of a multiclustered architecture for the specific application under consideration before implementing it. To show the possible results that are achievable with the extended power model, we will consider some case studies based on the optimization of some multimedia applications on the Lx architecture.

The specification of the Lx architecture allows us to divide the blocks of the processor in two types of blocks: blocks that are in common among all the clusters (*shared blocks*, such as the instruction cache), and blocks that are related to a single cluster (*non shared blocks*), i.e., their number varies linearly with the number of clusters. Here, we assume that the shared blocks power consumption does not increase with the number of clusters; this assumption results in a model whose forecasted power consumption is a lower bound of the actual power consumption, but is enough accurate to perform relative estimates.

To extend the model of Eq. 5.3, we introduce three additional parameters:

- C: the number of clusters
- $P_{ns}, P_s = (1 - P_{ns})$: percentage of power consumption related to blocks that are non shared (ns) or shared (s) in the multi-cluster architecture

Th P_{ns}, P_s are estimated on the monocluster architecture by considering the power consumption of the non-shared and shared blocks over the total power consumption. These two parameters can be composed into a *scaling factor* $\lambda(C) = (P_{ns}C + P_s)$ that multiplies the coefficients $U(\mathbf{0}|\mathbf{0}), M$ and S.

To simplify the computation of the model, an average value of the function $\nu(w_n^k|w_{n-1}^k) = \overline{\nu}$ can be used. $\overline{\nu}$ is weighted with the average

$ILP(C)$ of the stream (i.e., the number of syllables per instruction). $ILP(C)$ depends, obviously, on the number of clusters and on the ability of the compiler to exploit the availability of clusters. Note that the contribution $\overline{\nu}$ is associated to the non-shared blocks of the system.

The model of Eq. 5.3 can be rewritten by taking into account the scaling factor $\lambda(C)$ and $ILP(C)$ in the following way:

$$E(\mathcal{W}, C) \approx N\Big[\lambda(C)U(\mathbf{0}|\mathbf{0}) + ILP(C)\overline{\nu} + \\ + \lambda(C)(m * p * S + l * q * M)\Big] \quad (5.8)$$

After a careful analysis of the specification of the Lx multicluster, we measured the percentage P_s and P_{ns}, obtaining the following percentage values:

$$P_s = 0.337 \quad (5.9)$$

$$P_{ns} = 0.663 \quad (5.10)$$

As we will show in the rest of this chapter, the model of Eq. 5.8 can be used to perform an early analysis of a specific embedded application, to assess the suitability of the multicluster implementation.

5.3.1 Multicluster Design Space Exploration

To perform the design space exploration, we use the model of Eq. 5.8 combined with a configurable power and performance model of the caches of the system derived from [122]. The parameters that depend on the execution trace of the program, that is the number of instructions N, the ILP and the cache miss probabilities are estimated by an experimental version of the Lx toolchain that is able to compile and simulate programs for the multi-cluster configuration.

The design space is characterized by the following sub-spaces:

- Number of clusters: 1, 2, 4
- I-cache dimensions: 2048, 4096, 8192, 16384, 32768, 65536 [byte]
- D-cache dimensions: 2048, 4096, 8192, 16384, 32768, 65536 [byte]

The other cache parameters have been set to the values of the single cluster configuration: I-cache (block = 64 Bytes, associativity = 1), D-cache (block = 32 Bytes, associativity = 4).

5.3.2 Energy-Delay product analysis

The first analysis that is performed is the search for an optimum configuration that minimizes the Energy-Delay product (i.e., the product between the total energy consumed by the system and the number of execution cycles). In this case, this is called *unconstrained* analysis since no constraint on the maximum delay or the maximum energy of the system are set. Usually, this analysis lends to a configuration whose energy and delay is a trade-off between the two.

The design space exploration has been performed on a set of 32 C applications among which a set of FIR filters, fast cosine transforms, AES encryption/decryption and a subset of the public domain *Mediabench* suite (ADPCM encode and decode, EPIC decode and encode, JPEG encode and decode, MPEG2 encode and decode).

In total, over the 32 applications analysed, we found that only 3 applications benefited from the multi-cluster configuration. The two most significant cases are illustrated in Figures 5.7 and 5.8. Figure 5.7 shows the "Adpcm Encoder" behavior of the Energy-Delay product when clusters and caches are varied over the specified range. As can be seen, it is not suitable to use a multi-cluster configuration, since the minimum energy-delay product, highlighted by a circle, is provided by a sigle cluster architecture. Since the energy computed by the model is a lower bound on the energy consumption of the processor, we can reasonably assume that the real multicluster implementation will follow the same trend.

On the other hand, Figure 5.8 shows that the "Fir1" benchmark reaches its minimum Energy-Delay product using a multi-clustered configuration with 2 clusters. In this case, a more accurate implementation of the processor, and the corresponding power model, is needed to assess the goodness of the multicluster implementation, since this represents only a lower bound on the processor energy consumption.

5.3.3 Energy-Delay Pareto analysis

The second analysis consists in the construction of the Pareto curves of the energy and delay metrics of the system. A two-dimensional Pareto curve is defined as the set of points that are not dominated in both the dimensions by any other point.

Pareto curves are often used to perform a costrained analysis of the design space. In fact, they are useful to assess the best configuration for a particular metric given a constraint on the other one(s).

Figures 5.9 and 5.10 show the Energy-Delay scatter plot and the corresponding Pareto curves for two significative benchmarks among the 32 used for the analysis. Figure 5.9 represents the scatter plot and

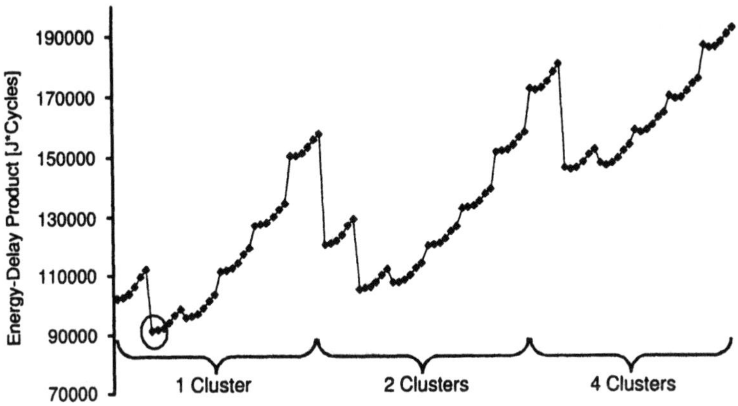

Figure 5.7. Energy-Delay product for the ADPCM encoder benchmark, by varying clusters and cache dimensions

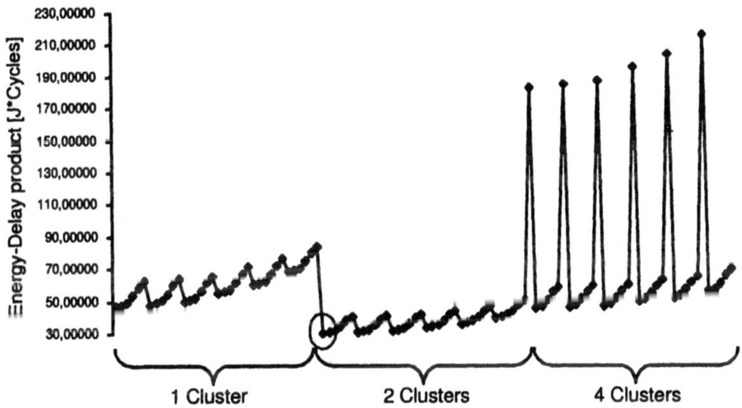

Figure 5.8. Energy-Delay product for the FIR1 benchmark, by varying clusters and cache dimensions

the Pareto curve for the MPEG2 decoder benchmark. As can be seen, the mono-cluster Pareto curve is the Pareto curve of the entire system. In other words, the mono-cluster configuration dominates the benefits (both in terms of performance and energy) of the multi-cluster configuration.

Figure 5.9 represents the scatter plot and the Pareto curve for the AES benchmark (encode and decode with a key of 128 bits). In this case the two-cluster configuration may be chosen if the performance constraint required by the designer is not satisfied by the mono-cluster configuration, which is the one that consumes less energy.

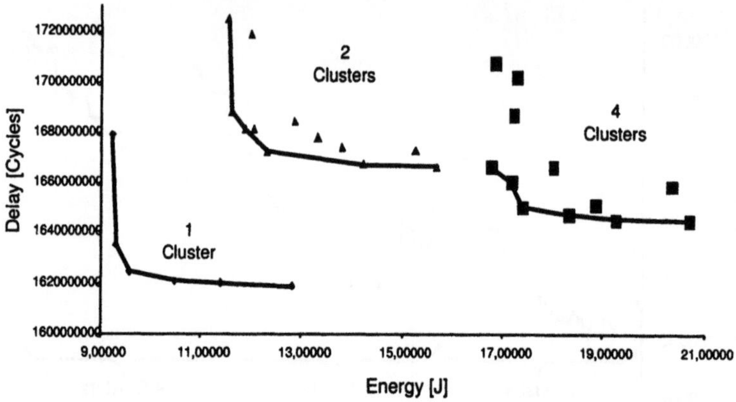

Figure 5.9. Energy-Delay scatter plot for the MPEG2 decoder benchmark, by varying clusters and cache dimensions

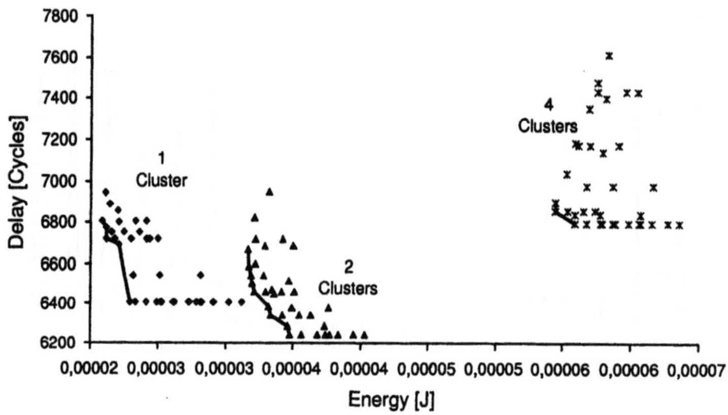

Figure 5.10. Energy-Delay scatter plot for the AES encoder and decoder benchmark, by varying clusters and cache dimensions

6. Conclusions

In this chapter, we have presented the application of the proposed instruction level energy model to a real industrial processor. It has been shown how a reduction of the complexity of the energy model for VLIW cores can be done by exploting the clustering the operations in the ISA based on the average energy behavior of the operations. Moreover, this chapter has shown how the model can be used to perform a low-power energy rescheduling and an early design exploration of a multi-clustere version of the Lx.

In the next chapter we show the plug-in of the model into an ISS and the consequent loss in accuracy due to the non-exact modeling of the behavior of the processor-lanes and cache misses.

The next chapter is devoted to the description of a framework for modeling and estimating the power consumption at the system-level for embedded architectures based on the Lx processor that is built upon the power model of the core presented here. The system-level simulation framework features dynamic power profiling capabilities during an ISS-based simulation, providing also a break-out of the power contributions due to the single components of the system (i.e., the core, the register file and the caches).

Chapter 6

SYSTEM-LEVEL POWER ESTIMATION FOR THE LX ARCHITECTURE

This chapter describes a technique to model and to estimate the power consumption at the system-level for embedded architectures based on the Lx VLIW processor. The method is based on a hierarchy of dynamic power estimation engines: from the instruction-level down to the gate/transistor-level. Power macro-models have been developed for the main components of the system: the VLIW core, the register file, the instruction and data caches. The previous chapter has introduced an instruction-level power model for the Lx processor core as well as the characterization and validation of the model based on the estimation of some architectural parameters, such as the probability per instruction per lane as well as data and instruction cache miss probabilities, by means of RT-level simulation. This chapter mainly extends the previous discussion by showing the plug-in of the Lx core power model into the ISS (Instruction Set Simulator) and the consequent loss in accuracy due to the non-exact processor behavior such as the behavior of processor's lanes and cache misses with respect to a RT-level simulation. Since our overall goal of the work is to define a system-level simulation framework for the dynamic profiling of the power behavior during the software execution, we further developed power models for the others parts of the system, mainly the register file and the caches. In this way, we can derive also a break-out of the power contributions due to the single components of the system. The experimental results, carried out over a set of benchmarks for embedded multimedia applications, have shown an average accuracy of 5% of the instruction-level estimation engine with respect to the RTL engine, with an average speed-up of four orders of magnitude.

V. Zaccaria et al. (eds.),
Power Estimation and Optimization Methodologies for VLIW - based Embedded Systems, 87–101.

1. Introduction

The request of low-power VLSI circuits for portable systems is steadily increasing. At the same time, CAD tools are evolving to deal with the complexity of systems-on-chip (SoCs) integrating one or more processors, the memory sub-system and other functional modules. The system-level design approach requires the effective management of large design complexity and the support of the specification and the analysis at high levels of abstraction. High-level power estimation and optimization is the crucial task in the early determination of the power budget for SoCs, for which accuracy and efficiency must be traded-off to meet the overall power and performance requirements.

In this scenario, our work focuses on software-level power estimation for embedded applications, where the embedded core and the memory sub-system are integrated in the SoC. The main contribution of our estimation approach consists of providing power consumption figures for software running on a given hardware architecture, and to help optimizing the target application for energy efficiency. The relative accuracy is certainly useful, while the absolute accuracy is the overall target.

State-of-the art power estimators for processors are based either on *instruction* [98] or *on micro-architectural* [123, 94, 99] modeling methodologies. The proposed approach aims at exploiting the main advantages of both techniques, however some differences can be remarked. If we consider instruction-level power analysis (ILPA), our approach gives better insight on the power bottlenecks during software execution (and optimization), because it is based on a more detailed micro-architectural model of the core.

The proposed system-level power estimation methodology is based on a hierarchy of dynamic power estimation engines at several abstraction levels: from the instruction-level down to the gate- and transistor-level. The main goal is twofold: from one side, to profile dynamically the power behavior during software execution and, from the other side, to provide the break-out of the power contributions due to the single components of the system. The proposed approach has been integrated into an industrial development framework, where a detailed description of the processor hardware architecture is available. In this chapter, we prove the viability of high-level power estimation for processor cores both from the *efficiency* and from the *accuracy* standpoints. The main contributions of our work can be summarized as follows:

- the development of power macro-models for the main components of the system: the VLIW core, the register file and the caches;

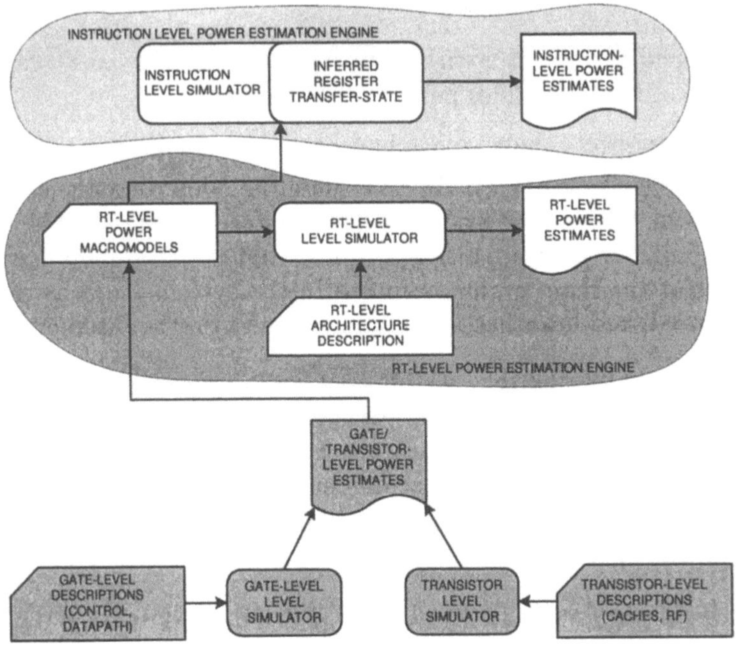

Figure 6.1. Power estimation framework.

- the development of a validation methodology to evaluate the accuracy of the macro-models against post-layout circuit and gate-level simulation;
- the integration of the power macro-models into a hierarchy of simulators, from RT-level (cycle-accurate) to the instruction-level.

The rest of the chapter is organized as follows. Section 2 describes the overall power estimation framework based on an instruction-level (IL) engine characterized by using an RT-level engine. The energy models for the VLIW core, the register file and the instruction and data caches are discussed in Section 3, while experimental results derived from the application of the proposed methodology to the Lx case study are described in Section 4. Finally, Section 5 outlines some concluding remarks and future developments of our research.

2. Power Estimation Framework

In this section, we present the proposed power estimation framework, based on a hierarchy of dynamic power estimation engines.

The cornerstone of our framework is the instruction-level power estimation (ILPE) module (see Figure 6.1). The ILPE module is based on an instruction set simulator (ISS) connected with a set of RT power macromodels and it is characterized by a very high speed: approximately 1.7 millions of bundles per second on average.

The ISS interprets the executable program by simulating and by profiling the effects of each instruction on the main components of the architectural state of the system (e.g., register and program counter values, state of the memory hierarchy etc.) and it generates a cumulative report of the main events occurred in the system (such as cache misses, mispredicted branches and other statistics on the instruction stream).

Eventually, the ISS can be modified to obtain a quick estimation of other RTL parameters (used as inputs for the RT power macro models) that would be otherwise not observed during a normal instruction set simulation. Among these parameters, we can find the Hamming distance between the consecutive values on the cache buses and register file input and output ports.

Instruction-level parameters as well as inferred RT-level parameters can be elaborated at *run-time* by plugging in the RT power model into the source code of the ISS or *off-line*, i.e., by post-processing the statistics of the output report of the ISS. Obviously, in the first case, the information is processed, averaged and represented instruction-by-instruction (not cycle-by-cycle) providing a rough idea of the instantaneous power consumption $P(t)$. In the second case (*off-line* elaboration or post-processing), the power consumption is evaluated at the end of the simulation as an average value of the function $P(t)$.

The accuracy of the IL power estimates depends on how well the ISS infers the correct RT-state and must be traded off with the ISS speed. Experimental results have shown an average accuracy of approximately 5% of the IL engine with respect to the RTL engine, while the performance improvement is of four orders of magnitude. As can be noted from Figure 6.1, the RTL power models are also embedded into a functional RTL description of the core, written in Verilog, that is used as reference for the IL engine.

The RTL power macro-models have been characterized by either gate-level analysis (with back-annotation of wiring capacitances extracted from layout) for synthesized modules, or by transistor-level power analysis for post-layout full-custom modules, such as cache memory banks and RF. All macro-models are linked to a cycle-accurate RTL simulation model of the core through the standard PLI interface. Thus, RTL power estimation is obtained as a by-product of RTL functional simulation.

2.1 Target System Architecture

The proposed framework has been applied to the scalable and customizable Lx processor technology [116] described in the previous chapter. The synthesizable RTL and the gate-level netlist of the processor have been used for the power measurements of the core module. The experiments have been carried out by using Synopsys VCS 5.2 and a set of PLI routines to generate toggle statistics over the gate-level netlist. Synopsys PowerCompiler has been used to combine the toggle statistics with the power models of the standard cells library provided by STMicrolectronics. Finally, the instruction and data caches as well as the register file power models have been characterized by simulating at the transistor-level the full-custom layout descriptions with an extensive set of input patterns.

3. Power Macro-Modeling

In this section, we present the macro-models developed to describe the power behavior of the main resources of the target system architecture, namely the VLIW core, the RF, and the separated I- and D-caches. The main issues of the proposed power macro-models can be summarized as follows:

- tightly relation with the micro-architectural details of each system module;
- accurate consideration of the processor-to-memory communication in terms of read/write accesses to each level of the memory hierarchy;
- availability at both RTL and IL to estimate the power consumption.

3.1 VLIW Core Model

The description of the analytical energy model used for the Lx and its validation process have been already presented in the previous chapter. In this section, we simply discuss the comparison between the estimates given by the model plugged into the ISS and the estimates gathered at the RTAs (see Table 6.1). The average error between the ISS-based estimations and the RTL-based estimations is approximately -12% while the average error is around zero. These results are particularly interesting, since they show the error introduced by an inaccurate estimation of the instruction-level parameters.

3.2 Register File Model

The general problem of estimating the power consumption of RFs has recently been addressed in [124], where different RF design techniques

	gauss	fir1	fir2	fast_dct	fast_idct	dct_idct
ISS Power [W]	0.468	0.528	0.519	0.401	0.438	0.486
RTL Power [W]	0.439	0.497	0.536	0.448	0.467	0.436
Percentage Error	6%	6%	-3%	-12%	-7%	10%

Table 6.1. **Comparison between IL power estimates and RTL power estimates for the benchmark set for the core processor**

have been compared in terms of energy consumption, as a function of architectural parameters such as the number of registers and the number of ports.

In this section, we propose a parametric power model of a multi-ported RF, in which we assume that the power behavior is linear with respect to the number of simultaneous read/write accesses performed on the different ports:

$$P_{RF} = P_i + \frac{1}{T} \sum_{1 \leq n \leq N} (E_{r,n} + E_{w,n})$$

where P_i is the RF base power cost measured when neither read nor write accesses are performed, T is the total simulation time, $E_{r,n}$ ($E_{w,n}$) is the energy consumption of a read (write) access occurred during bundle w_n, and f_S has been defined above.

The energy contribution $E_{r,n}$ is defined as:

$$E_{r,n} = \sum_{1 \leq i \leq N_{rp}} H(RR_{i,n}, RR_{i,n-1}) * E_{rb}$$

where N_{rp} is the number of read ports of the RF, H is the Hamming distance function, $RR_{i,k}$ is the data value read from the RF output port i by the k-th bundle and E_{rb} is the energy consumption associated with a single bit change on a read port.

The energy contribution $E_{w,n}$ is defined as:

$$E_{w,n} = \sum_{1 \leq i \leq N_{wp}} H(RW_{i,n}, old_{i,n}) * E_{wb}$$

where N_{wp} is the number of write ports of the RF, H is the Hamming distance function, $RW_{i,n}$ is the new data value written by the n-th bundle on input port i, $old_{i,n}$ is the previous data value contained in the same RF location and E_{wb} is the energy consumption associated with a single bit change on a write port.

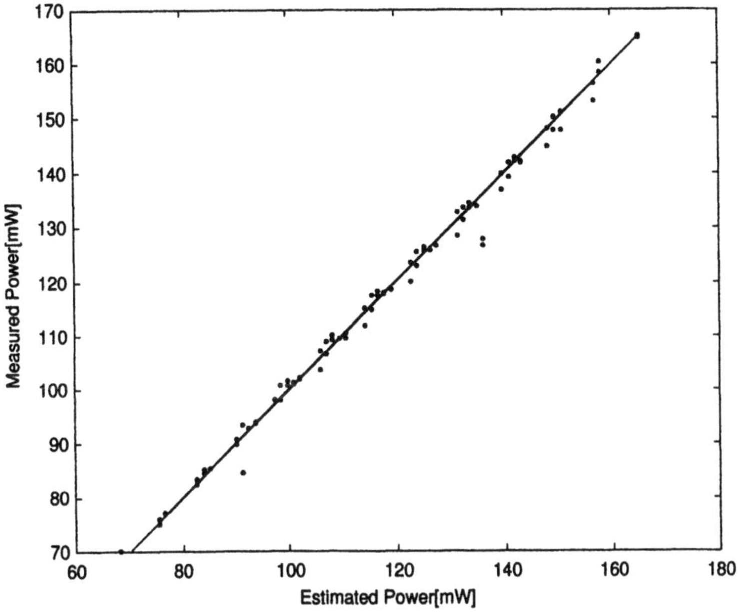

Figure 6.2. Agreement between measured transistor-level power values and estimated power values for the Register File (Maximum error within ±8%)

3.2.1 RT-level Model Validation

The model characterization and validation has been carried out by a transistor-level simulation of the register file circuit extracted from layout including parasitics. The simulation has been performed by generating a set of sequences of input vectors characterized by different switching activity on the read and write ports.

The agreement between estimated and measured power values is shown in Fig. 6.2 (maximum error 8%, RMS 1.75%.)

3.2.2 ISS Model Validation

Similarly to the core model validation, the accuracy of the RT-level and IL simulations for the Register File have been compared. The results in Table 6.2 reports the accuracy obtained by the ISS power model compared to RTL power model. The maximum error is approximately -27 %, while the average error is approximately 4%. The maximum error can be due to the fact that the ISS does not infer the switching activity associated with the accesses to the specific register file ports, but only an average value depending on the type of instructions. In particular, the maximum error found can be considered acceptable in our target

	gauss	fir1	fir2	fast_dct	fast_idct	dct_idct
ISS Power [W]	0.088	0.115	0.121	0.075	0.071	0.09
RTL Power [W]	0.089	0.098	0.115	0.087	0.09	0.09
Percentage Error	-1%	15%	5%	-16%	-27%	0%

Table 6.2. Comparison between IL and RTL power estimates for the benchmark set for the Register File

system architecture since, as we can see in Section 4, the contribution of the Register File to the overall system power is maintained within 5%.

3.3 Cache Model

Most published analytic cache models [125] deal with relatively simple cache organizations, and they are not suitable for modeling complex cache architectures based on multiple SRAM memory banks, with a significant amount of control logic. The multi-banked structure is dictated mainly by performance constraints, since cache access time is critical for overall processor performance.

The modeling approach adopted in this chapter is the hierarchical approach described in [126, 127]: first, we built power macro-models for all the different types of SRAM banks in the caches, second we compose these models in a single logical model that generates the correct access patterns for every bank according to the cache organization. Composition of the atomic macro-models in the complete cache model is trivial at the RT level, because the RTL description of the cache subsystem contains the behavioral description of every SRAM module. Deriving the cache model for IL simulation is not as straightforward, because the ISS simply provides cache accesses per instructions, but it does not infer any knowledge of internal cache organization.

3.3.1 RTL power models for caches

The macro-models for the atomic SRAM modules are *mode-based:* power consumption depends on the mode of operation (i.e., `read`, `write`, `idle`). More precisely, since the SRAM modules are synchronous, the energy consumed in a given clock cycle is mainly a function of the *mode transition* between the previous and the current cycle. Thus, we characterized energy as a function of the nine possible mode transitions (e.g., `read-read`, `read-write`, etc.). For a given mode transition, energy is weakly dependent on the number of transitions on the address lines. Accounting for this dependency leads to a macro-models with $9 \cdot (N_{addr} + 1)$ characterization coefficients, where N_{addr} is the number of address lines. Thus, we can model the energy consumed by cache modules with the following equations [126, 127]:

$$E_{tag} = E_s N_s + \sum_{n=0}^{8} \sum_{x=s,r,w} \sum_{y=r,w} E_{xy}[n] N_{xy}[n] \tag{6.1}$$

$$E_{Imem} = E_s N_s + \sum_{n=0}^{10} \sum_{x=s,r,w} \sum_{y=r,w} E_{xy}[n] N_{xy}[n] \tag{6.2}$$

$$E_{Dmem} = E_s N_s + \sum_{n=0}^{10} \sum_{x=s,r,w} \{E_{xr}[n] N_{xr}[n] + \sum_{B=0}^{8} E_{xw}[B,n] N_{xw}[B,n]\} \tag{6.3}$$

where E_s is the energy consumed during an idle cycle for the selected module, N_s is the number of idle cycles during the entire RTL simulation, the parameter $E_{xy}[n]$ is the energy consumed during one access operation depending on the number $[n]$ of bits that change, the current cycle operation y and the access operation x, performed during the previous cycle.

For the D-cache, E_{Dmem} contains an additional parameter:

$$\sum_{B=0}^{8} E_{xw}[B,n] N_{xw}[B,n] \tag{6.4}$$

since the D-cache line can be written byte per byte with a bus composed of byte selection signals. In such case, the term $E_{xw}[B,n]$ depends on both the previous access address n and the current byte selection B.

This type of model can be efficiently implemented into the Verilog description, with a Look-Up Energy table. Each row of this table can be selected with an access code that depends on the parameters x, y and n previously described. The coefficients has been characterized by simulating the back-annotated transistor-level netlist of the SRAM modules with the MACH-PA circuit simulator by Mentor Graphics. Average accuracy of the SRAM macro-models can be considered satisfactory (percentage average errors are within 5%), as shown in the next sub-section.

3.3.2 RT-level Model Validation

The scatter plot of Figure 6.3 represents the energy consumption evaluated at the RT-level compared to the energy measured with MachPA. Since an electrical simulation of the entire processor is impractical, for

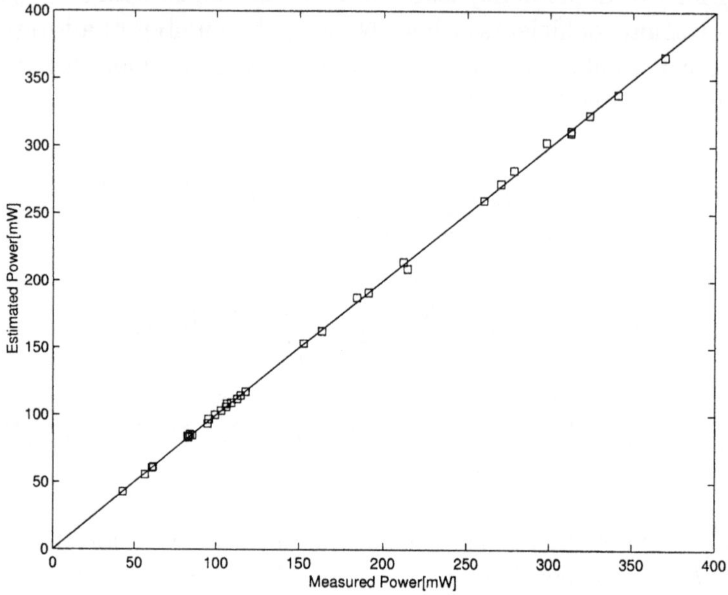

Figure 6.3. Agreement between measured transistor-level power values and estimated RT-level values for the cache blocks.

the validation we use the same decomposition approach used in the characterization phase. The scatter plot of Figure 6.3 shows the agreement between the RT-level and the transistor-level power measurements.

3.3.3 ISS Level Power Models for Caches

The ISS profiles the accesses to the cache resources.

In the case of a miss the cache is refilled with the behavior represented in Table 6.3.

To model the power consumption of I-cache, we consider the average energy consumed by the module for a given type of access. Since the ISS is instruction-accurate, we do not have the visibility of access addresses on a cycle-by-cycle basis, as in the RTL model. For this reason, we use the average values associated with the given type of access (`sleep`, `read`, `write`) (gathered from the Look Up Energy table).

The power consumption in D-cache is computed in a similar fashion. The main difference is the capability to perform writes in 64-bit blocks during a cache refill, and the different topology of the architecture that is 4-way set associative. In this case, we distinguish only between read or write accesses. The behavior of the data cache is summarized in Table 6.4.

Cycle	ram_x	ram_y	tag_x	tag_y
1-33	read	idle	read	read
34	2 bank write, 2 bank read	idle	read	read
35	read	idle	read	read
36	2 bank write, 2 bank read	idle	read	read
37	read	idle	read	read
38	2 bank write, 2 bank read	idle	read	read
39	read	idle	read	read
40	2 bank write, 2 bank read	idle	read	read
41-45	read	idle	read	read
46	2 bank write, 2 bank read	idle	read	read
47	read	idle	read	read
48	2 bank write, 2 bank read	idle	read	read
49	read	idle	read	read
50	2 bank write, 2 bank read	idle	read	read
51	read	idle	read	read
52	2 bank write, 2 bank read	idle	write	read

Table 6.3. Refill behavior of the I-cache

Event	Cycle	Tag	Bank	Comment
Read hit	1	4 tags read	8 banks read	
Write hit	1	4 tags read	1 banks read	
Read miss	1	4 tags read	8 banks read	(such as read hit)
	2	4 tags read	2 banks read	dirty bit extract
	4	/	4 banks write (64bit switching)	refill action
	5	1 tag write	/	TAG write
	6	4 tag read	8 bank read	pending read action
Write miss	1	4 tags read	/	
	2	4 tags read	2 banks read	dirty bit extract
	4	/	4 banks write (64bit switching)	refill action
	5	1 tag write	/	TAG write
	6	4 tag read	8 bank read	pending write action
	7	/	1 bank write	pending write action

Table 6.4. Behavior of the D-cache

3.3.4 ISS Model Validation

Table 6.5 reports the accuracy of the ISS power model compared with the RTL power model. The maximum error in the estimation D-cache

	gauss	fir1	fir2	fast_dct	fast_idct	dct_idct
I-cache ISS Power [W]	0.635	0.640	0.636	0.626	0.618	0.635
I-cache RTL Power [W]	0.614	0.619	0.617	0.589	0.602	0.617
Percentage Error	3.4%	3.4%	3.1%	6.3%	2.7%	2.9%
D-cache ISS Power [W]	0.430	0.766	1.05	0.314	0.284	0.398
D-cache RTL Power [W]	0.429	0.774	0.905	0.266	0.265	0.433
Percentage Error	0.2 %	-1.0 %	16.0 %	18.0%	7.2%	-8.1%

Table 6.5. Comparison between Instruction-Level power estimates and RTL power estimates for the benchmark set for the caches

power consumption is approximately 18 %. This value is due to the fact that the D-cache has a very complex behavior compared to the I-cache and it is more difficult to gather the parameters needed for the model.

4. Experimental Results

In this section, we show some experimental results obtained by the instruction-level power estimation engine proposed in this work. The IL engine is based on the ISS available in Lx toolchain, that has been modified to gather a fast estimate of the RTL status and parameters. These values have been linked to the power macro-models to get power estimates. The experiments have been carried out over a set of selected benchmark applications including *C* language implementation of digital filters, discrete cosine transforms, etc., especially tuned for the Lx processors.

Fig. 6.4 shows the comparison results between the IL power estimates with respect to RTL estimates based on the same models. For the benchmark set, the average error (IL vs. RTL) is 5.2%, while the maximum error is 7.9%. On a Sun Ultra Sparc 60 at 450MHz (1GB RAM), the RTL engine simulates 160 bundles per second on average, while the IL engine simulates 1.7 millions of bundles per second on average, thus providing a speed-up of four orders of magnitude approximately.

Figure 6.5 shows an example of the ISS-based static power profiling for the same benchmarks of above. As can be seen from the figure, the power dissipation associated with the caches is more than a half of the total power consumption. The contribution of the D-cache is however highly dependent on the benchmark, while the I-cache power consumption is almost stable. This is due to the fact that the I-cache is accessed at every clock cycle, even during an I-cache miss, while the D-cache is not enabled during inactivity periods.

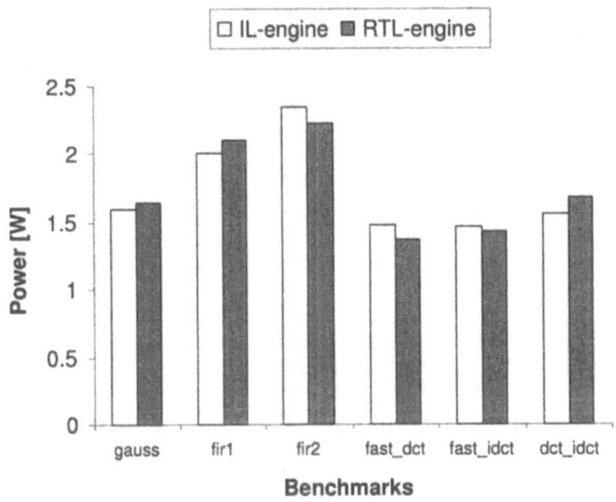

Figure 6.4. Comparison between Instruction-Level power estimates and RTL power estimates for the benchmark set. (Average Error 5.2% - Maximum Error 7.9%).

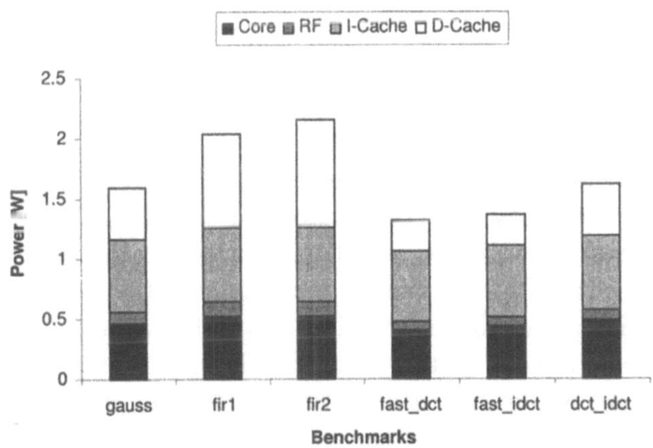

Figure 6.5. Break-out of the RTL power contributions due to core, RF, I-cache, and D-cache for the benchmark set.

Figure 6.6 shows an example of ISS-based dynamic power profiling applied to a FIR filter. During the first 600 ns, there is a high total power consumption that is due to an intensive D-cache activity (in this phase the data is fetched from the memory hierarchy into the registers). The remaining part of the plot shows a period in which the data are elaborated and only the core, the RF and the i-cache consume power.

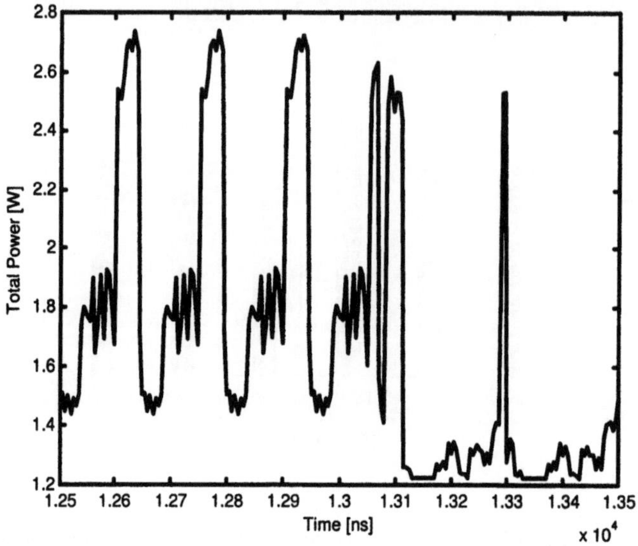

Figure 6.6. An example of ISS-based system-level dynamic power profiling capability.

5. Conclusions

In this chapter, we have presented an efficient and accurate framework for embedded core power modeling and estimation. The method is based on a hierarchy of dynamic power estimation engines: from the instruction-level down to the gate/transistor-level. Power macro-models have been developed for the main components of the system: the VLIW core, the register file, the instruction and data caches. The main goal consisted of defining a system-level simulation framework for the dynamic profiling of the power behavior during the software execution, providing also a break-out of the power contributions due to the single components of the system. Experimental results, carried out on a Lx-based system during the execution of a set of benchmarks for embedded multimedia applications, have demonstrated an average accuracy of 5% of the instruction-level estimation engine with respect to the RTL engine, with an average speed-up of four orders of magnitude. Future directions of our work aim at defining power efficient instruction scheduling opportunities and techniques to optimize the software code from the power standpoint.

This chapter concludes the discussion of the novel methodologies for power estimation presented in this book. The next chapters analyze the problem of power optimization in VLIW-based architectures and introduce the two power optimization methodologies presented in this book.

II

POWER OPTIMIZATION METHODS

Chapter 7

BACKGROUND

Chapter 3 presented the overall power consumption sources in a digital microprocessors and, in particular, the four parameters on which a designer can operate to reduce the power consumption: voltage, physical capacitance, switching activity and frequency (see Equation 3.8). All the power reduction techniques presented in this chapter represent an attempt to reduce one of these parameters. In the following sections, we briefly analyze the power optimization techniques available at each level of abstraction in the representation of a microprocessor, starting from technology optimizations up to system-level optimizations. Concerning the system-level optimizations, we focus on the state-of-the-art design space exploration methodologies as well as dynamic power management techniques since they are of fundamental importance in the context of this work.

1. Technology/physical-level optimizations

Several optimizations can be performed directly on the technological parameters as well as on physical-level description without modifying other high-level characteristics of the system. As can be seen from Equation 3.8, a clock frequency decrease causes a proportional decrease in the power dissipation. However, this approach is not viable for high-performance digital systems, where the clock frequency is one of the primary design targets to do not violate peak-performance throughput constraints. In fact, reducing the clock frequency results in a slower computation, the power consumption over a given period of time is reduced, but the total amount of work per unit of time reduced as well. In other words, the energy dissipated to complete the given task does not change. Under the assumption that the total amount of energy pro-

V. Zaccaria et al. (eds.),
Power Estimation and Optimization Methodologies for VLIW - based Embedded Systems, 105–121.

vided to complete the task is constant, decreasing the clock frequency only increases the time needed to complete the given task [128, 62].

Furthermore, for portable battery-operated systems, the total amount of energy provided by the batteries is not a constant, but it depends on the rate of discharge of the battery. In fact, if we decrease the discharge current, we can increase the total amount of energy drawn from the battery. In other words, there can be some advantages in reducing the clock frequency, since the batteries are more proficiently used when the discharge current is small.

Voltage scaling is the most effective way to reduce the switching power, since $P_{switching}$ depends quadratically from the power supply voltage [109]. The main drawback of voltage scaling techniques is the consequent increment in the delay of signal propagation through the logic gates. A simple model of the delay associated with a CMOS inverter is given by the following expression:

$$T_d = \frac{C_L \times V_{dd}}{I} = \frac{C_L \times V_{dd}}{\frac{\mu C_{ox}}{2}(W/L)(V_{dd} - V_t)^2} \tag{7.1}$$

where μC_{ox} is a technology-dependent constant, W and L are respectively the transistors width and length, and V_t is the threshold voltage. The most important assumptions behind the previous equation are: i) the current through the MOS transistor is well fitted by the quadratic model, ii) during the transient, the device controlling the charge (discharge) of the output capacitance is in saturation.

When all the other parameters are fixed, the inverse delay shows an inverse proportionality dependence on the supply voltage (see Figure 7.1). As can be noted, the speed decreases dramatically as V_{dd} gets closer to the threshold voltage V_t. This is due to the low conductivity of the channel when V_{gs} approaches V_t and it can be avoided by lowering also V_t. However there are some limits in reducing the threshold voltage manly due to decreasing noise margins and increasing sub-threshold currents I_{ds}. Noise margins may be relaxed in some cases of low power designs because of the reduced currents being switched, however the sub-threshold currents can result in significant leakage power dissipation.

More in general, the main limitation of voltage scaling approaches is that they assume the designer has the freedom of choosing the appropriate voltage supply for the design. Unfortunately, the introduction of accurate voltage converters are expensive, and the management of multiple supply voltages can contribute to complicate board- and system-level designs resulting in an increase of the overall system cost. Thus, in many cases, to control the supply voltage arbitrarily can be unfeasible.

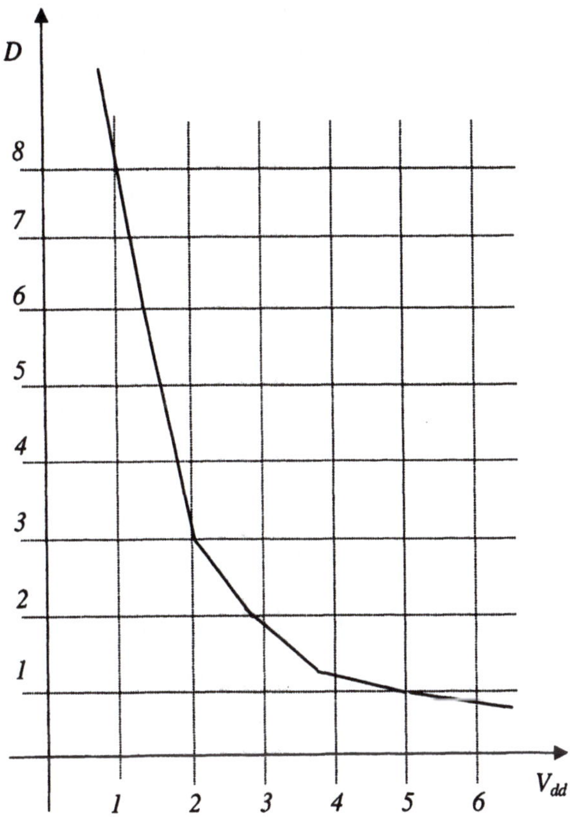

Figure 7.1. Normalized delay (D) versus the supply voltage (V_{dd} [V]) for a typical gate in a standard CMOS process.

One of the most important products that implements dynamic frequency and voltage scaling is the Intel Pentium III featuring Intel's SpeedStep technology. SpeedStep involves three components: the processor, a compliant BIOS that recognizes the use of battery or AC power, and a motherboard chip that regulates voltage and speed (SpeedStep controller). The SpeedStep controller is used by the OS to define the new state by adjusting the voltage regulator and the CPU frequency multiplier. All of these operations takes less then 1ms. Other commercial processors such as AMD Athlon/Duron as well as Transmeta's Crusoe exploit similar mechanism to reduce power consumption.

Another technique for power minimization is capacitance reduction: power dissipation is dependent on the physical capacitances seen by individual gates in the circuit. Capacitances are reduced by using less logic, smaller devices, and fewer and shorter wires within the digital circuit, thus resource sharing, logic minimization, and gate sizing can be seen as techniques for low power consumption. However, while reducing device sizes reduces physical capacitance, it also reduces the current driven by the transistors, increasing the delay of the circuit. When a netlist of gates is transformed into the geometric specification (*layout*) of the system, some optimization can be done to decrease the load on high switching activity gates by proper netlist partitioning, placement, gate and wire sizing, transistor sizing and reordering, and, finally, routing.

Particular care must be introduced concerning the clock signal, which is the fastest and most heavily loaded net in a digital system. Ideally, clock signals should have minimum rise/fall times, specified duty cycles, and zero skew. A large fraction of the power consumption of a digital system is consumed by the clock net, thus the total capacitive load seen by the clock source must be minimized. This can be done by introducing a chain of drivers at the source by extending or sizing wires for zero-skew or by inserting buffers to partition a large clock tree into a small number of subtrees with minimum wire widths [129].

2. Logic-level and RT-level optimizations

Circuit and logic-level optimization are the next level of abstraction at which a system can be optimized. In this section, we show the main techniques proposed in the literature so far to reduce power consumption at these levels of abstraction.

2.1 Circuit Factorization

Two-level Boolean function is often implemented to have additional levels between inputs and outputs. Multilevel factorization of a set of such Boolean functions provides a reduction in area [130] and it has been extended in [131, 132] to compute the goodness of a divisor by taking into account also possible power savings.

The process of factoring out common subexpressions is split into two algorithms: *kernel-extraction* and *substitution*. The kernel extraction algorithm finds multiple or single-cube divisors and retains divisors that are common to two or other functions. The best common divisors (in terms of power savings) are factored out and the affected functions are then simplified by means of the substitution algorithm.

2.2 Low power technology mapping

Technology mapping consists of binding a set of logic equations to the gates in some target cell library. To facilitate mapping, the logic circuit is decomposed into a set of base functions such as two-input NAND/NOR gates. A minimum area mapping algorithm has been proposed in [133] and implemented in tools such as DAGON and MIS. The algorithm reduces the technology mapping problem to DAG covering problem and approximates DAG covering by a sequence of tree coverings performed optimally by means of dynamic programming. More specifically, a tree is a useful structure that can express circuits in which each gate has a fanout of one while a DAG can be expressed by one or more eventually overlapping trees. Power consumption can be also reduced during the technology mapping phase [134, 135, 136]. Tiwary *et al.* [135] propose a tree covering algorithm starting from a root node and attempting to match a library cell that minimizes a power cost function. The cost function takes into account the cost of the match (in terms of intrinsic and output capacitances as well as output switching activity) an the minimum cost of the input nodes of the match (computed by calling the algorithm recursively).

Generally, the power-delay mapper reduces the number of high switching activity nets at the expense of increasing the number of low switching activity nets. In addition, it reduces the average load on the nets. In this way the total weighted switching activity and hence the total power consumption in the circuit is minimized. This procedure leads to an average of 18% reduction in power consumption at the expense of 16% increase in area without any degradation in performance.

2.3 Gated clocks

Clock gating is a technique for selectively blocking the clock signal to a flip-flop or a register, and forcing the circuit to do not make transitions when an idle condition is detected.

At the logic-level, the concept to minimize the useless switching activity by disabling the clock signal has been manually applied by digital designers for a long time. However, automatic dynamic power management techniques for single registers in a sequential block has been proposed only recently [137, 138, 139] and consists of automatically synthesizing the Boolean function that represents the idle conditions of the circuit. Figure 7.2a represents a finite state machine whose next state depends on the inputs of the finite state machine and on the state itself. It turns out that when the next state is equal to the current state, the flip flop is unnecessarily triggered by the clock, consuming useful power. Figure 7.2b shows a power-aware FSM in which the clock is conditioned to an activation signal generated by the module F_a. The activation signal is first latched by a transparent latch L (that blocks the signal when the clock is high to prevent glitches) and then is "and"-ed with the original clock. The module F_a generates a high value by watching on the current state of the finite state machine and on the next state; if they match the clock is disabled.

2.4 Gate freezing

Gate freezing [140] is an optimization technique for the minimization of the glitch power. In this technique, some existing gates of the circuit are replaced with functionally equivalent gates (called F-Gates) that can be "frozen" (i.e., not propagate glitches) by means of an opportune control signal. The overhead for the generation of control signals is reduced by means of clustering of gates that try to minimize the amount of logic needed.

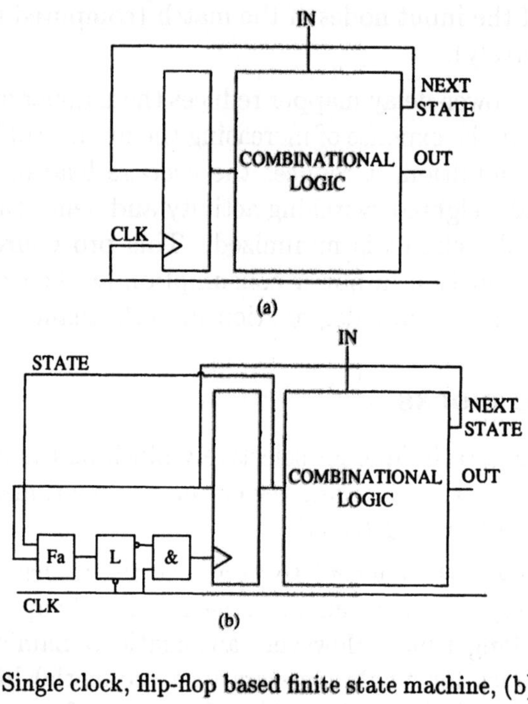

Figure 7.2. (a) Single clock, flip-flop based finite state machine, (b) clock gated finite state machine

The drawback of this approach is that the redundant logic circuit can offset the power savings achieved by the clock management as well as negatively impacting the control of the clock skews. Testing and verification can be also affected and some attempts have been made to design testable gated clock circuits [141].

2.5 Pre-computation

Pre-computation techniques [142] consist of minimizing the activity within a combinational module that computes a generic function

$f(R1(t), R2(t))$, where $R1(t) = x(t-1)$ and $R2(t) = y(t-1)$ (see Figure 7.3). The technique consists of predicting at clock cycle $t-1$ the value $f(R1(t), R2(t))$ with two predictor modules $g_1(x(t-1))$ and $g_2(x(t-1))$ where:

$$g_1(x(t-1)) = 1 \rightarrow f(R1(t), -) = 0 \tag{7.2}$$

$$g_2(x(t-1)) = 1 \rightarrow f(R1(t), -) = 1 \tag{7.3}$$

where "-" stands for whatever value of $R2(t)$. If either g_1 or g_2 is 1 then register $R2$ is clock gated avoiding unuseful activity within the module. Since g_1 and g_2 implement only a part of the functionality of the combinational module, they are much more smaller and can work with a subset of the inputs.

An example that illustrates the precomputation logic is the n-bit comparator that compares two n-bit numbers C and D and computes the function $C > D$. Assuming that each $C[i]$ and $D[i]$ has a 0.5 signal probability, the probability of correctly predicting the output results using the most significant bit is 0.5 regardless of n. Thus, a power reduction of 50% (by ignoring the hardware overhead of the control logic, g1 and g2) can be achieved.

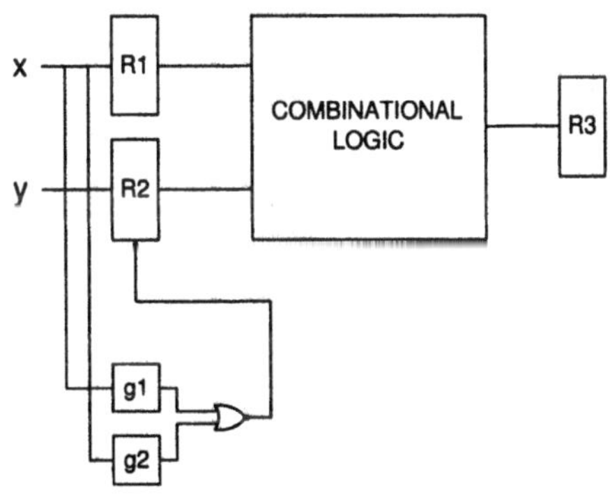

Figure 7.3. The precomputation architecture

2.6 Finite state machines for low power

Finite state machines are widely adopted to design the control unit of a modern microprocessors. FSMs are usually described by means of state transition graphs (STG) that represent a form *behavioral* description. An STG is then transformed in an RT-level form by assigning a binary code to each of the states and by generating the corresponding *output* and *next state* Boolean functions.

To minimize the power consumption, a common approach is to compute the transition probability of a given edge of the STG and to transform the problem into the embedding of the state transition graph into a hypercube so that edges with high transition probability are associated with the states with low Hamming distance. This combinatorial optimization problem can be solved with standard or heuristic search techniques [143, 144, 145, 146]while the computation of the state probabilities can be carried out either exactly or by using approximate techniques [147].

2.7 Retiming

Retiming techniques consists of changing the position of registers within a circuit to perform the same operations in different clock cycles [148]. In [149] retiming has been used to reposition latches within a circuit in order to reduce the propagation of spurious activity. Registers are positioned iteratively by taking into account a cost function that incorporates the number of latches, to avoid an excessive increase in area.

3. Microarchitecture level optimizations

Microarchitectural power reduction techniques are usually coupled with technology-level transformations to obtain circuits with reduced energy dissipation and acceptable performance (see Section 1). A common approach is first to increase the performance of the design and then reduce the voltage as much as possible. Proposed methods to do this include increasing algorithm concurrency and pipelining by using faster units, and increasing the number of hardware units used.

Among the microarchitectural level optimization techniques presented in literature, *parallelization* and *pipelining* are the most important. Parallelization consists of increasing the number of operations performed in each clock cycle, by adding more function units to the processor, and decreasing the clock period and the supply voltage. In [109] it is shown that, when doubling the resource of a data-path, power savings of up 2.15X can be obtained with substantially no performance overhead.

Pipelining consists of inserting latches within the processor datapath in order to decrease the value of the critical path. The resulting positive slack on timing constraints is exploited by maintaining the original clock frequency but reducing the supply voltage [109].

Speed-up techniques such as these typically result in increased silicon area and therefore are commonly referred to as "trading area for power".

Varying the size of the system structures is another way of optimizing the power consumption of a processor without affecting considerably the performance. Among the structures that can be configured, we can find the various buffers used for instruction fetching and reordering as well as branch prediction buffers. Typically, such type of analysis consists of a trade-off analysis between the power and the energy consumed by the system and the actual performance and is measured in terms of Power Delay Product, Energy-Delay Product (EDP) [150] and Energy-Throughput Ratio (ETR) [128]. Such metrics are often modified to focus more on energy (E^2D, E^3D etc..) or on delay (for high end systems).

MIPS/Watt Ratio [151] (and more recently SPEC/Watt ratio [92])is the inverse of the Power Delay Product and is another way of evaluating power and performance trade-offs. Also in this case, there are some derived metrics useful to find trade-offs focused on performance ($MIPS^2$/Watt, $MIPS^3$/Watt) and power.

Wattch [94] and SimplePower [93] are two architectural simulators of superscalar processors that have been extensively used to explore such type of optimizations. System-level design space exploration (described in Section 5) are based on such type of micro-architectural analysis, but include also cache memory exploration and software level exploration.

Another way of optimizing power consumption at the microarchitectural level is to use some compression schemes for the data/control word flowing through the pipeline. Among these techinques, Canal *et al.* [152] proposed a pipeline data encoding scheme in which data, addresses, and instructions are compressed by maintaining only significant bytes with two or three extension bits appended to indicate the significant byte positions. The extension bits flow down the pipeline and enable pipeline operations only for the significant bytes. This technique has shown a reduction in switching activity of up to 40%.

Researchers at IBM [153] have applied *distributed decode* and *microcode* to reduce power consumption. In this scheme, each opcode that flows through the pipeline becomes a function of the previous op-code and can be set to 0, 1 or the previous value. By changing only the signals that are required for that function, the power of the machine can be reduced. For example a NOP operation would look similar to both an ADD and a SUB operation with the signal RegisterEnable=OFF.

Other authors propose to skip the fetch and decode stage of the pipeline by using a small RAM (Decoded Instruction Buffer) to store decoded instructions [154]. Power savings have been shown of up to 40%.

In [155] a methodology for automatically determining an assignment of instruction op-codes that minimizes the switching activity inside the registers of the pipeline stages is proposed. To guide the assignment of the binary patterns to the opcodes, profiling information on instruction adjacency collected during ISS simulation is used. The effectiveness of the methodology has been proved on a MIPS R4000 RISC microprocessor.

3.1 Bus Power Reduction

The load capacitances of the off-chip I/O signals and the width of internal data and control busses are usually significant. Therefore remarkable power savings can be achieved by applying bus encoding techniques to reduce the switching activity.

The Bus-Invert code [156] is a low-power encoding scheme for data patterns that works as follows: the Hamming distance between two successive bus values is computed and, if it is larger than $N/2$ (N=bus width) the current value is negated and then transmitted; otherwise, it is transmitted as is. A redundant bus line INV is needed to transimt the bus polarity to the receiving end. This method guarantees a maximum of $N/2$ transitions per clock cycle, and it performs well when the values to be transmitted are randomly distributed in time and no information about their correlation is available (appropriate for data-bus encoding).

Since instruction addresses are mostly sequential, Gray coding [157] has been proposed to minimize the transitions on the instruction address bus. The Gray code ensures that when the data is sequential, there is only one transition between two consecutive data words. This algorithm is optimum for sequential data only in the class of irredundant codes.

The T0 code [158], is another low power redundant bus encoding technique which requires an extra line (called INC) to signal that the current address is consecutive to the previous one. When INC is high, the current bus value is frozen to avoid switching activity, and the new address is computed directly by the receiver. When two addresses are not consecutive, the INC line is low, and the bus operates as usual. Several variants of the T0 code have been proposed; in [159] a T0 variant that incorporates the Bus-Invert principle (exploiting distinctive spectral characteristics of the streams) has been proposed. The T0-XOR code, [160], is an irredundant version of the T0 code in which the *xor* function is used to decorrelate the values sent on the bus and eliminate the INC line.

The working zone code [161] is based on the observation that many programs access several data arrays. The accesses to each array are generally in sequence, but they are often interleaved, decreasing the sequentiality on the bus. The working-zone scheme restores sequentiality by storing the reference addresses of each working zone on the receiver side and by sending only the highly sequential offsets. When a new working zone is accessed, this information is communicated to the receiver with a special code on the bus. The receiver changes the default reference address, and offset transmission can resume. The problem of this approach is that it still relies on strong assumptions on the patterns in the stream and that it requires one extra bus wire for communicating a working-zone change.

The Beach code [162, 163] relies on the consideration that there can exist some temporal correlations, other than arithmetic sequentiality, between the patterns that are being transmitted over the address bus. The basic idea consists of determining an encoding strategy that depends on the particular address stream being transmitted. The technique groups the bus lines into clusters according to their correlations and an encoding function is automatically generated for each cluster. Clearly, the computation of the encoding functions is strictly dependent on the execution trace and works at best on special-purpose systems, where a dedicated processor (e.g., core, DSP, microcontroller) repeatedly executes the same portion of embedded code.

Finally, Benini *et al.* [164] present algorithms for the automatic synthesis of encoding and decoding interfaces, minimizing the number of transitions on bus lines by constructing low-transition activity codes, encoders and decoders. An adaptive architecture, adjusting encoding to reduce switching activity, is also introduced.

4. Instruction-level optimizations

Instruction-level or *software*-level optimizations refer to a class of source and machine code transformations whose effect is to reduce power during the execution of the code itself. A first attempt to reduce power consumption by means of code scheduling can be found in [99], where it is applied to a DSP architecture with *instruction packing*. The proposed algorithm schedules one basic block of a DSP program at a time, by taking into account the overhead cost associated with the various pairs of instructions. The first phase of the algorithm works on a data flow graph (DFG) of each basic block, by applying an ASAP packing procedure to pack instructions. The second phase attempts to construct a schedule for the packed instructions while minimizing the total overhead cost by means of a list scheduling algorithm [165].

Su *et al.* [166] proposed a technique in which Gray code addressing and *cold scheduling* techniques are combined. The cold scheduling algorithm is a list-based algorithm in which the priority is determined by the power cost associated with the pairs of instructions. This approach has shown a reduction of up to 30% in the control path of the processor.

Toburen *et. al* [167] propose a scheduling approach based on a cycle-by-cycle energy threshold that can not be violated in any given cycle. Instructions are scheduled with a list-scheduling algorithm until the threshold for a given cycle is reached. Significant energy savings have been reported with negligible performance impact.

Chang and Pedram [168] describe a technique for transforming the low power register assignment problem into a minimum cost clique covering of an appropriately defined compatibility graph. A compatiblity graph is a graph where vertices correspond to data values, and and a directed edge is present between two verices u and v if and only if their life times do not overlap and the u comes before v. The switching activity between pairs of edges that could potentially share the same register is used as a cost to find a clique with minimum cost covering the graph. The problem is solved optimally (polynomial time) using a max-cost algorithm.

Mehta *et al.* [169] propose to reduce power consumption by proper register re-labeling, i.e., by encoding the register labels such that the sum of the switching activity costs between all the register labels in the transition graph is minimized. These techniques have led to a maximum energy reduction of 9.82%.

Kandemir *et al.* [170] investigated several state-of-the-art compiler optimizations on several benchmarks (among which loop unrolling, fusion, cache block tiling etc.) They found that the energy consumption of the memory system is higher than the core while executing unoptimized code and that most code optimizations increase the power consumption of the core.

Kandemir *et al.* [171] also investigated several instruction scheduling algorithms for VLIW architectures, that reorder a given sequence of instructions taking into account the energy considerations. They propose also scheduling algorithms that consider energy and performance at the same time. The results obtained have shown that these techniques are quite successful in reducing the energy consumption while preserving partially the original performance.

In [172] a power optimization methodology based on preprocessing and restructuring the code at the source level has been presented. Other techniques such as horizontal and vertical scheduling have been proposed by Lee *et al.* [117]. Horizontal scheduling tries to directly minimize the switching activity on the instruction bus by choosing suitable pairs of instructions through a bipartite matching scheme. Since the vertical scheduling algorithm is an *NP*-hard problem, it is heuristically solved by reducing it to a bipartite matching scheme on a limited instructions window. Parikh *et al.* [173] modified list-scheduling by trading-off energy and speed at the same time.

Finally, a number of approaches have been proposed in the literature dealing with the code optimization to reduce the memory accesses and data placement in the memory hierarchy in such a way to minimize power consumption [174, 175, 176, 177, 178, 179].

5. Design space exploration at the system level

Many multimedia applications implemented today on portable embedded systems have stringent real-time constraints together with the necessity of minimizing power consumption (e.g., portable MP3 decoders and digital cameras). The increasing complexity of today's embedded systems is forcing the designers to determine an optimal system structure as early as possible during the design process due to the tight time-to-market requirements. The selection of the most important modules of the system and their conformance with respect to the application and the possible alternatives must be performed at the highest levels of abstraction. In fact, it is impossible in terms of design time to synthesize and to analyze each part of the design space, even considering the optimization of only two parameters such as power and performance. Furthermore, at the highest abstraction levels, several degrees of freedom can be exploited to achieve a comprehensive design optimization, while maintaining a good level of accuracy of the estimated parameters. Accuracy and efficiency must be traded-off to meet the system-level overall energy and delay requirements, avoiding costly re-design phases.

Several system-level estimation and exploration methods have been recently proposed in literature targeting power-performance trade-offs from the system-level standpoint. Among these works, the most significant methods can be divided into two main categories: *(i)* system-level power estimation and exploration in general, and *(ii)* power estimation and exploration focusing on cache memories.

In the first category, the SimplePower approach [93] can be considered one of the first efforts to evaluate the different contributions to the energy budget at the system-level. As cited above, the SimplePower exploration framework includes a transition-sensitive, cycle-accurate datapath energy model that interfaces with analytical and transition sensitive energy models for the memory and bus sub-systems. The SimpleScalar estimation framework has been designed to evaluate the effects of some high-level algorithmic, architectural and compilation trade-offs on energy. Although the proposed system-level framework is quite general, the exploration methodology reported in [123] is limited to a search over the space of the following parameters: cache size, block buffering, isolated sense amplifiers, pulsed word lines and eight different compilation optimizations (such as loop unrolling).

The *Avalanche* framework presented in [180] simultaneously evaluates the energy-performance tradeoffs for software, memory and hardware for embedded systems. The approach is based on system-level and takes into consideration the interdependencies between the various system modules. The target system features the CPU, D-cache, I-cache and main memory, while the impact of buses is not accounted for. For this reason, the estimates are less accurate in advanced submicron technologies, for which routing capacitances play a significant role in the system's power consumption.

In [181], a system-level technique is proposed to find optimal configurations for low-power high-performance superscalar processors tailored to specific user applications. More recently, the Wattch architectural-level framework has been proposed in [94] to analyze power vs. performance trade-offs with a good level of accuracy with respect to lower-level estimation approaches. Low-power design optimization techniques for high-performance processors have been also investigated in [182, 183] from the architectural and compiler standpoints.

A trade-off analysis of power/performance effects of SOC (System-On-Chip) architectures has been recently presented in [184], where the authors propose a simulation-based approach to configure the parameters related to the caches and the buses. The approach is viable only when it is possible to build a model either analytically or statistically (with a high level of accuracy) for the energy and delay behavior of the system's modules. Moreover the reported results are limited to finite ranges of the following parameters: bus width, bus encoding, cache size, cache line size and set associativity.

For what concerns the power exploration of the cache hierarchy, the authors of [150] propose to sacrifice some performance to save power by filtering memory references through a small cache placed close to the processor (namely *filter cache*). A similar idea has been exploited in [185], where memory locations with the highest access frequencies are mapped onto a small, low-energy, and application-specific memory that is placed close to the core processor.

A model to evaluate the power/performance trade-offs in cache design has been proposed in [186], where the authors discuss also the effectiveness of novel cache design techniques targeted for low-power (such as vertical and horizontal cache partitioning). An analytical power model for several cache structures has been proposed in [125]. The model accounts for technological parameters (such as capacitances and power supplies) as well as architectural factors (such as block size, set associativity and capacity). The process models are based on measurements reported in [187] for a 0.8 μm process technology.

The previous analytical model of energy consumption for the memory hierarchy has been extended in [188] and [189] where the cache energy model is included in a more general approach for the exploration of memory parameters for low-power embedded systems. The authors consider as performance metrics the number of processor cycles and the energy consumption and show how these metrics can be affected by some cache parameters such as cache size, cache line size, set associativity and tiling size, and off-chip data organizations.

Benini *et al* [190] propose a novel methodology to map the most frequently accessed data onto a small on-chip memory (called ASM). The memory architecture is based on an ad-hoc decoder that maps processor addresses onto ASM addresses. Experimental results collected on a MP3 decoder have shown significative energy savings over the same architecture in which ASM is not used.

6. System level power management

The power consumption of modern electronic systems can be reduced dynamically by means of suitable power management policies [191]. A power management policy decides when to change the operating conditions of a resource (typically *on* and *off*) on the basis of the system history and the workload applied.

A power manageable system is a system consisting of a set of power consuming resources (referred to as *service providers*), a power manager (regulating the power consumption of the resources) and a power management policy (defining how the power manager regulates power consumption).

The power manager handles stimuli arriving from the environment (modeled as a *service requester*) and issues the corresponding commands to system resources to change their power states. Environmental stimuli consist of a series of requests arriving either directly from users (e.g., requests to read or write the disk) or from other devices such as the network interfaces. The power manager processes these requests to forecast the next power state for each system resource.

Possible policies can range from very simple ones, like keeping everything "on" all the time (maximum power consumption) or turning off a resource as soon as it becomes idle, to more realistic and efficient ones that take into account trade-offs between power and performance by automatically adapting to actual workloads. An optimal policy would shut down the disk during an idle period T whenever the energy spent during a complete transition ($on \rightarrow off \rightarrow on$) is less than the energy spent by the resource kept in the *on* state. Given a resource whose power in the on state is P_{on} and whose transition $on \rightarrow off \rightarrow on$ is characterized by an energy consumption E_{tr}, the minimum time T for which the following inequality holds:

$$P_{on} \times T > E_{tr}$$

is called *break-even* time T_{be} and it can be considered as the minimum idle period for which it is convenient to shut down that resource.

Predictive techniques are the most common power management policies used in modern electronic systems. These power management schemes are based on the prediction of idle periods of a resource. If the predicted idle time T is greater than or equal to T_{be}, the power manager shuts down the resource.

Predictive techniques try to exploit the correlation between the past history of the workload and its future behavior in order to predict future events efficiently. Possible prediction errors can be:

- over-estimates: an idle time longer than the actual one is forecast.
- under-estimates: an idle time shorter than the actual one is forecast.

Several metrics have been used to assess the goodness of a predictor in the power management area [192]. The *safety* of a predictor is defined as the probability that the predictor does not make an over-prediction. The *efficiency* is defined as the probability that the predictor does not make under-predictions. An ideal predictor has safety and efficiency equal to one. The implementation of an ideal predictor for actual workloads is very difficult to obtain.

The fixed timeout [193] policy is probably the most popular predictive technique. When an idle period begins, the power manager observes its duration and, if it is greater than a predefined timeout threshold T_{to}, the PM issues an *off* command to the resource. The system remains off until a request from the environment is sent to the resource.

The problem of timeout strategies is that a significant amount of power is wasted before the timeout expiration. Predictive shutdown policies [194, 195, 196, 197, 198] try to solve this problem by predicting the duration of an idle period. If the predicted value is greater than T_{be}, then the system is shut down instantaneously. Adaptation can be also be introduced to tune the policy parameters dynamically (see [194] for an application).

Threshold-based [195, 199] policies represent a variation of predictive shutdown policies. In these policies, the duration of the current busy period (or some derived parameter) is observed and if it is lower than a threshold time T_{th}, the next idle time period is assumed to be greater than T_{be}. In these policies it is assumed that short active periods are followed by long idle periods.

Predictive approaches assume that the type of transitions and the response times of the system are deterministic. However, modern digital systems are characterized by increasing complexity and modeling uncertainty, making the deterministic approach a hardly feasible solution. Stochastic/dynamic control techniques [200, 201] have been proposed to cope with these problems. In [200], the power manager chooses which command to issue to the power manageable system by modeling it as a unified controllable Markov chain. The commands are issued in probabilistic terms that is, each decision is characterized by a probability to be taken. This decision can be dependent either on the entire history of the system or only on the current system state. If the decision depends only on the current system state and this dependence does not change with time, the policy is defined as a stationary Markov policy. In [201] only the workload is modeled with a probabilistic state machine while the resource under consideration is modeled deterministically.

In most cases, these types of techniques lead to stationary policies. Some techniques have been proposed to overcome these limitations [202] exploiting on-line algorithms that use sliding windows to capture the non-stationarity of the system transition probabilities.

Chapter 8

A MICRO-ARCHITECTURAL OPTIMIZATION FOR LOW POWER

Estimation of power consumption in an architecture envisioned for low-power applications is, fairly often, the first step towards power-oriented design optimization. We present now a low-power approach to the design of embedded VLIW processor architectures that exploits the "forwarding" (also called "bypassing") hardware already existing in the Data Path and providing operands from inter-stage pipeline registers directly to the inputs of the function units. This hardware is here used for avoiding the power cost of writing/reading short-lived variables to/from the Register File (RF). In many embedded systems applications, a significant number of variables are short-lived, that is their liveness (from first definition to last use) spans only few instructions. Values of short-lived variables can then be accessed directly through the forwarding registers, avoiding write-back to the RF by the producer instruction, and successive read from the RF by the consumer instruction. The compiler (through its static scheduling algorithm) decides on enabling the RF write-back phase. Our approach implies a minimal overhead on the complexity of the processor's control logic and no significant increase in length of the critical path.

Application of the proposed solution to a VLIW embedded core, when accessing the Register File, has shown a power saving up to 26.1% with respect to the unoptimized approach on a set of target benchmarks. We conclude this chapter outlining possible extensions of the proposed approach that introduce a new level on the memory hierarchy, consisting of the pipeline microregisters level. This extensions virtually increase RF space and, thus, reduce the necessity of register spilling.

V. Zaccaria et al. (eds.),
Power Estimation and Optimization Methodologies for VLIW - based Embedded Systems, 123–141.

1. Introduction

Low-power design techniques are widely adopted during microprocessor design to meet the stringent power constraints present in many embedded systems [109] [63]. Such techniques apply either at the *technological* level or at the *logic and architectural* levels. In the first case, very widely adopted solutions involve voltage reduction and (selective) frequency reductions. This type of approach is quite general - i.e., application-independent - and is in fact used also where CPUs for laptops or PDAs are concerned.

For high-performance processors, such as VLIW, most low power approaches target on the reduction of the effective switched capacitance of each of the processor gates. The effective switched capacitance of a gate is defined as the product of the load capacitance at the output of the gate and its output switching activity.

While reducing the effective switched capacitance is suitable for the combinational and in-core sequential part of the processor, custom modules need special techniques for power reduction. Multi-ported register files represent the majority of such processor modules (apart from instruction and data caches), as they consume a substantial portion of the power budget.

Zyuban *et al.* [124] compare several optimization techniques aiming at the reduction of power consumption by register files. They propose a model for the register file energy consumption in each clock cycle that is a weighted sum of the energy of a read and a write operation:

$$E_{cycle}(t) = n_w(t) * E_{write} + n_r(t) * E_{read} \qquad (8.1)$$

where $E_{cycle}(t)$ is the register file energy consumption during clock cycle t, $n_w(t)$ ($n_r(t)$) is the number of write (read) accesses to the RF during clock cycle t, E_{write} (E_{read}) is the energy consumption per write (read) access.

Thus, energy consumption is modeled both by parameters that depend on the particular application software ($n_w(t), n_r(t)$) and by circuit/architectural level parameters (E_{write}, E_{read}). While various techniques exist for reducing the circuit- and architectural-level parameters (see [124] for a comprehensive overview), parameters deriving from characteristics of the application software have not been directly addressed in the literature dealing with low-power design.

In this chapter, we propose a power optimization technique based on the reduction of the number of reads and write accesses per clock cycles by exploiting the *forwarding* network of a VLIW processor. Such optimization can be applied whenever in the object code there are short-lived variables whose value can be retrieved from the forwarding network without being written back to (or read from) the register file.

2. Background

Forwarding (also called *bypassing* and sometimes *short-circuiting*) is a solution adopted in pipelined microprocessors to reduce performance penalties due to data hazards (Read-after-Write conflicts) [51]. The forwarding hardware bypasses the results of previous instructions from the inter-stage pipeline registers directly back to the function units that require them. To implement the bypassing mechanism, the necessary forwarding paths and the related control must be included in the processor design [51, 203]. In general, this technique requires a bus connecting the output of any pipeline register following the EXE stage(s) to the inputs of any function unit present in the EXE stage(s). Data bypassed from an instruction to some function unit in earlier pipeline stages are then stored in the Register File (RF) when the instruction reaches the last pipeline stage (i.e. the write-back stage).

In general, the forwarding paths for VLIW processors present an implementation complexity that can impact on the processor's critical path. In [204], such design issues are discussed and different bypassing interconnection networks which can be used for increased levels of connectivity among the functional units of the processor are presented. This analysis deals with both frequency of stalls and cycle time penalties; it shows that a bypassing network completely connecting all the function units does not provide a performance improvement when the cycle time penalty and the area of the chip are taken into account.

The concept of taking advantage of register values that are bypassed during pipeline stages has been combined with the introduction of a small register cache in [205] to improve performances. In this architecture ('Register Scoreboard and Cache') the register file is replaced with a register cache and a suitable backing store. If the normal bypass mechanism supplies a sufficient fraction of all register operands, the register cache can be very small. Such cache can reduce the critical path delay due to the register file of typical processors, thus enabling frequency improvements.

For superscalar processors, a scheme including an analysis at compilation time and an architecture extension to avoid the useless commits of short-lived variables has been proposed in [206]. The advantages provided by this approach have been evaluated by the authors, mainly in terms of reducing the number of write ports for the RF and decreasing the amount of spill code needed, thus improving execution time. Per-

Cycle	Instruction
I_n	$\$r2 \leftarrow \ldots;$
I_{n+1}	$\ldots;$
I_{n+2}	$\ldots \leftarrow \$r2;$
I_{n+3}	$\$r2 \leftarrow \ldots;$

Figure 8.1. Fragment of code showing a short-lived variable

formance improvements of this solution have been also reported; on the contrary, the impact on power consumption has not been considered.

The concept of dead value information (DVI) in the RF has been introduced in [207] where "dead" register values can be exploited to aggressively reclaim physical registers used for renaming architectural registers. This allows using a smaller and faster physical register file and potentially increasing the processor's clock rate . The authors propose also using DVI to eliminate unnecessary save/restore instructions from the execution stream in the case of procedure calls and across context switches.

3. Main contribution of this work

The optimization technique here presented exploits data forwarding for short-lived variables to save power in VLIW pipelined architectures. The basic idea consists of exploiting the inter-stage registers and forwarding paths to reduce the RF activity and avoid power-expensive writing of short-lived variables, thus reducing RF activity. Short-lived variables are simply stored by the producer instruction in the inter-stage registers and fed to the consumer instruction by exploiting the forwarding paths. No write-back to the RF is performed by the producer instruction, and no read from the RF is affected by the consumer instruction.

As an application example, let us consider the fragment of code in Figure 8.1. In this example, the write-back in I_n of the results into register $r2 can be avoided and the successive use of the same variable in I_{n+2} can be performed directly by accessing the variable from the forwarding network since the $r2 liveness is equal to two instructions.

In our approach, the decision of enabling the RF write-back phase is anticipated at compile time by the compiler static scheduler. This solution limits the modification ot the processor's control logic: only simple additional decoding logic (operating in parallel with the normal instruction decode unit) is required, and the critical path is increased by a one-gate delay to control the write enable signal of the RF.

To apply this optimization in the case of a generic variable to be stored in register R, the compiler must compute the *liveness length* L of the n-th assignment to R, defined as the distance between its n-th assignment and its last use. (We denote here a variable by the name of the register in which it is supposed to be written). This information enables the compiler to decide whether the variable must be actually stored in the RF for further use or whether its use is in fact limited within few clock cycles. In the latter case, the variable is *short-lived*, and its value can be passed as an operand to subsequent instructions by using the forwarding paths, thus avoiding to write it back to the RF.

The proposed architecture becomes particularly attractive in some classes of embedded applications, where the register liveness analysis (see Section 6) has shown that the greatest part of register definitions have a liveness limited to the next two instructions.

In the next sections, we shall analyze the impact of the proposed low-power architectural solution for VLIW processors on the ISA and on the compiler. The impact (if any) of the solution on the exception handling mechanisms will be discussed (5); the concept of microregisters will then be introduced.

4. Low-Power Forwarding Architecture

In order to analyze the hardware support needed by the proposed low power optimization technique, we refer to the 4-way VLIW processor architecture with 5-stage pipeline already examined in the previous sections and provided with forwarding logic (see Figure 8.2). The pipeline stages are:

- IF: Instruction Fetch from I-cache.
- ID: Instruction Decode and operands read from RF.
- EX: Instruction Execution in one-cycle latency ALUs.
- MEM: Load/Store memory accesses.
- WB: Write-Back of operands in the RF.

There are three forwarding paths (EX-EX, MEM-EX and MEM-ID) provide direct connections between pairs of stages through the EX/MEM and MEM/WB inter-stage registers. Given the forwarding network of Figure 8.2 and considering a sequence $W = w_1 \ldots w_k \ldots w_n$ of very long instruction words, a generic instruction w_k can read its operands from the following instructions:

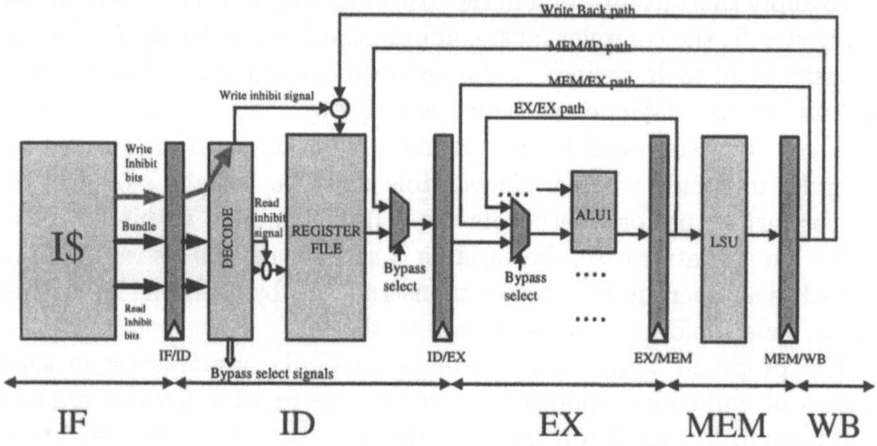

Figure 8.2. The proposed low-power 5-stage pipelined VLIW forwarding architecture.

- w_{k-1} through the EX/EX forwarding path (used when w_k is in the EX stage).
- w_{k-2} through the MEM/EX forwarding path (used when w_k is in the EX stage).
- w_{k-3} through the MEM/ID forwarding path (used when w_k is in the ID stage).
- w_{k-n} where $n > 3$ through the *RF*.

The solution proposed to optimize power consumption requires dedicated logic in the ID stage to decide whether or not the source/destination registers must be read-accessed in the RF. The instruction format must encode dedicated Read/Write Inhibit bits to enable/disable the corresponding accesses to the register file. The Write Inhibit bit of the instruction format is decoded to deassert the Write Enable signal (see Figure 8.2) in the RF write port. The Read Inhibit bit is used to keep unchanged the values on the input read addresses of the RF. This action reduces the switching activity of the read ports in the RF. Globally, the hardware overhead is equal to one-gate delay added on the processor critical path.

From the above, it follows that for each destination (source) register there must be a Write (respectively, Read) Inhibit bit that shall be used to avoid activity on the RF. To grant availability of such bits, we propose three different approaches:

1 Add specific operation bits in the long instruction encoding format. This solution has to be adopted at the time of ISA definition, and it requires increasing the length of instruction encoding. For each operation in the bundle we add one bit for each source (destination) register to indicate whether it is a read (write) inhibited register or not. These bits are used to de-assert write enable (read enable) lines in the write-back (respectively, instruction decode) stage, an action that requires using an AND gate for each read/write port of the register file. With this solution, for example, a bundle composed of four ternary operations (one destination and two source registers) would require twelve additional bits. In turn, this would involve increasing the word length of I-cache (and of main memory) as well as the width of data buses accessing such memories (the alternative, i.e., keeping a 32-bit operation length, would lead to halving the number of addressable registers). All this would increase both costs and power consumption; this last point should be accounted for during the evaluation of power savings.

2 Exploit unused bits in instruction encoding. This solution is suitable when the ISA has been already defined; instruction length is kept unchanged, but application of the proposed approach is restricted only to the subset of ISA operations characterized by unused bits in the instruction format. In general, unused operation bits are very few. If we assume one unused bit per operation, we can consider exploiting it to avoid useless writes only. Note that, in this case, there is no power overhead at all.

3 Finally, a possible alternative (that does not require any modification to the instruction format) requires departing from the strict "static" approach typical of "pure" VLIW architectures to introduce some measure of dynamic analysis of the object code. This leads to implementing by means of additional hardware the control logic for the Read (Write) Inhibit bit. Given the limited instruction window available to the control logic for analysis of variable liveness, we must restrict the analysis to checking Write-after-Write dependencies; in fact, the last Read on a register can be dynamically checked by relatively simple hardware only by monitoring two consecutive Writes to the same register. On our in-order five-stage pipeline, the window for such analysis corresponds to the sequence of instructions between the Decode stage and the Write Back stage; thus, Write-after-Write dependencies involve instructions separated by three cycles at most. Increasing instruction length (as in solution 1) and limiting the set of instructions for which the technique is applicable (as in solution

2) are both avoided; on the other hand, the power overhead due to the added control logic and the limited instruction window examined negatively contribute to the final power saving.

Coming now to the required software support, we consider only the first two solutions, keeping the strictly static scheduling approach typical of pure VLIW architectures. All scheduling decisions concerning data, resource and control dependencies are solved at compile time during static instruction scheduling [208]; thus, the decision whether the destination (source) register must be write (read) inhibited or not, has to be charged to the compiler, limiting the hardware overhead.

To set specific write/read inhibit bits at compile time, the compiler must compute the liveness length of each assignment to any register and compare such length with the limit implied by the pipeline structure in order to exploit the forwarding network.

The liveness length L of the n-th assignment to a register R is here defined as the distance (measured as the number of instructions) between the n-th assignment and its last use:

$$L_n(R) = U_n(R) - D_n(R) \tag{8.2}$$

where $D_n(R)$ is the clock cycle associated with the fetch phase of the instruction that performed the n-th assignment to R and $U_n(R)$ is the clock cycle associated with the fetch phase of the last instruction that used the n-th assignment to R before the redefinition of R during its $(n+1)$-th assignment $D_{n+1}(R)$.

In order to simplify the analysis that the compiler must perform, we assume a liveness length $L_n(R)$ to be valid when the following conditions hold:

- U_n and D_n belong to the same basic block;
- D_{n+1} and D_n belong to the same basic block.

These rules force us to consider only liveness ranges that do not cross the boundaries of basic blocks. This assumption, apparently very restrictive, does not actually represent a major concern, since most modern VLIW compilers maximize the size of basic blocks, thus generating a relevant number of intra-basic block liveness ranges.

The concept now introduced is best illustrated by means of an example. To this purpose, we analyze a portion of a 4-way VLIW assembly trace executing a DCT (see Figure 8.3). The code consists of four long instructions (instructions whose fetch phases are associated with clock

```
27268   shr  $r16 =  $r16, 8
        sub  $r18 =  $r18, $r7
        add  $r17 =  $r17, $r19
        sub  $r19 =  $r19, $r15;

27269   shr  $r18 =  $r18, 8
        shr  $r17 =  $r17, 8
        shr  $r19 =  $r19, 8
        mul  $r20 =  $r20, 181;

27270   sub  $r10 =  $r10, $r8
        mul  $r11 =  $r11, 3784
        sub  $r5  =  $r12, $r9;

27271   sub  $r10 =  $r10, $r3
        add  $r20 =  $r20, 128
        brf  $r26, label_232;
```

Figure 8.3. Example of a segment of a 4-way VLIW assembly execution trace of a DCT algorithm

cycles 27268, 27269, 27270, and 27271, respectively); each long instruction is from now on identified by the its fetch clock cycle. A long instruction is made up of a set of one to four operations, whose list is terminated by a semicolon. In this example, there is a boundary terminating a basic block at instruction 27271 (a conditional branch operation). Let us evaluate the liveness of the assignment of $\$r18$ in 27268 (D_n). This definition is used for the last time in instruction 27269, since there is another definition of $\$r18$ in the same cycle (i.e., D_{n+1}). Thus, L_n of $\$r18$ is equal to one clock cycle. It is not possible to compute the liveness length L_{n+1} of $\$r18$ because there are neither last uses U_{n+1} nor redefinitions D_{n+2} in the same basic block. Referring to the architecture of Figure 8.2, one can see that the maximum liveness that can be exploited, given the architecture of the pipeline, is 3. Thus, the write inhibit bit of the assignment to $\$r18$ performed in instruction 27268 can be set to true by the compiler, while for the assignment to $\$r18$ performed in instruction 27269 the write inhibit bit must be set to *false*.

Obviously, processors with more than five pipeline stages may have more than three forwarding paths, so that the application of the proposed low-power optimization can be extended to variables whose liveness length is higher than 3.

5. Exception Handling and Cache misses

An obvious critical point of the proposed optimization technique can be constituted by handling of exceptions and of cache misses. In such cases, results produced by instructions that precede the excepting one must be granted to be available in a "permament" memory (i.e., a register in the RF or a memory word) so that execution of the interrupted program may be correctly resumed after exception handling. Write-Back inhibiting could create problems in this context; we will now prove that correct execution can be easily granted, both for exact and for inexact exception handling.

For our subsequent analysis, the processor's state is seen as structured as follows:

1 A *permanent* architectural state stored in the RF.

2 A *volatile* architectural state stored in the registers between the pipeline stages from which the forwarding network retrieves source operands.

The volatile architectural state is handled as a FIFO memory, the depth being equal to the number of stages that keep stored in one of the pipeline registers the result of an operation (in the proposed 5-stage pipeline architecture the depth is equal to 3).

As a general rule, the organization of a pipelined processor grants that an operation's result present in the volatile state is automatically written back in the RF. On the contrary, given the pipeline management optimized for low-power presented above, whenever an element exits the volatile state and it is no longer of use, it can be discarded avoiding Write-Back in the RF.

In the reference architecture here adopted, exceptions can occur during the ID, EX, or MEM stages, and are serviced in the WB stage.

According to the exception taxonomy in [203], we assume that the processor adopts the precise mode exception handling mechanism. Under this assumption, exceptions can be *exact* or *inexact*. An *exact* exception, caused by an instruction w, makes the architectural state changes visible to all instructions issued after w. Furthermore, all state changes due to instructions issued before w are visible to the exception handler. When an *inexact* exception occurs, the instructions in the pipeline are executed until completion, and the exception is serviced afterward. In this case, instructions issued immediately after the excepting instruction will not see architectural state changes due to the action of the exception handler.

Let us analyze the behavior of the proposed solution in the two cases of exact and inexact exception handling. Assume first that exceptions are handled in *exact* mode. When the excepting instruction reaches the WB stage, instructions following it in the pipeline are flushed and re-executed. Refer to the example shown in Figure 8.4, where at cycle x an instruction w_k reads its source operands from a write-inhibited w_{k-2} instruction through the forwarding network. Let instruction w_{k-1} generate an exception during the MEM stage. The results of w_{k-2}, due to the setting of the write-inhibit bit, would have been lost; since these operands need to be used during the re-execution of w_k after exception servicing, the low-power optimization approach would cause errors in execution of the program. In fact, when the interrupted program will be resumed, neither the forwarding network nor the RF will contain the results produced by w_{k-2}, and the architectural state seen during re-execution of w_k (at cycle $x + nn$) will be incorrect.

To guarantee that instructions in the pipeline following the excepting one will be re-executed in the correct processor state, write-inhibited values must be written in the RF *anytime an exception signal is generated in the ID, EX or MEM stages.*

In the previous example (where w_{k-1} generates an exception in the MEM stage), the proposed solution forces the write-back of the results of $w_{k-1}$and w_{k-2} in the RF; during re-execution of w_k at cycle $x + nn$, the operands will then be read from the RF.

Let us now assume that exceptions are handled in *inexact* mode. In this case, to guarantee a semantically correct execution, all instructions in the pipeline must be forced to write back the results in the RF.

The proposed low-power architecture shown in Figure 8.2 supports both exception handling mechanisms with the following provisions:

- When the exceptions are *exactly* handled, the supported register liveness is less than or equal to 2 clock cycles (through the EX/EX and the MEM/EX paths);

- When exceptions are *inexactly* handled, the exploitable register liveness can be extended to 3 clock cycles (through the EX/EX, MEM/EX and MEM/ID paths).

Let us finally consider the case of interrupts and cache misses. Due to the asynchronous nature of interrupts, they can be treated as inexact exceptions by forcing each long instruction in the pipeline to write back its results before the interrupt handling routine is executed. As for instruction-cache misses, these produce bubbles that flow through the pipeline. Therefore, whenever a miss signal is raised by the I-cache

	Pipeline Stage		
Cycle	EX	MEM	WB
x	w_k	w_{k-1} (Exc. Signaled)	w_{k-2}
x+1	....	w_k	w_{k-1} (Exc. Served)
x+2	Deleted	Deleted	Deleted
....		....	
....	Exc. Handler	Exc. Handler	Exc. Handler
....		....	
x+nn	w_k	Exc. Handler	Exc. Handler

Figure 8.4. An example of exception handling.

control logic, we force the write back of the results of the instructions in the pipeline. (Concerning data cache misses, we assume that they stall the pipeline until the needed block is ready). Given this solution, the normal write inhibition mechanism can be tolerated since the relative timing of the instructions is not modified.

6. An Application Example

To provide experimental evidence of the viability of our approach, we refer once more to the Lx family of embedded VLIW cores and of the related software toolchain [116]. The six pipeline stages of Lx are: Instruction Fetch (IF), Instruction Decode (ID), Register Read (RR), Execution 1 (EX1), Execution 2 (EX2), Write-Back (WB). The forwarding paths are EX1-EX1 and EX2-EX1. The EX2-RR path is the normal write-back path; there are thus in total three paths that can be exlpouted for our low-power optimization approach. The Lx ISA is a RISC integer instruction set that supports speculative execution and prefetching. The RF provides sixty-four 32-bit general purpose registers and eight 1-bit branch registers.

6.1 Results of Register Liveness Analysis

An experimental environment has been set up to evaluate the impact of the proposed low-power optimization on the Lx architecture.The first relevant step consists obviously in evaluating the liveness length of registers. To this purpose, a suitable set of benchmark programs is needed; given the charateristics of the core and the foreseeable relevant applications, we focused on a set of multimedia benchmarks and embedded real-world DSP algorithms written in C. Most of these algorithms were taken from the Mediabench suite [121] while the remaining ones algorithms were in-house optimized versions of classical transforms (see

[209, 210] for an overview). Our main goal is to measure the dynamic percentage of register definitions in the application code that can be directly read from the forwarding network, without being written in the RF.

Register liveness analysis could be performed by the compiler on the static program; on the contrary, we decided to perform a dynamic analysis, consisting of the inspection of an execution trace of the program. This provides us with accurate run-time profiling information on how many register accesses can be inhibited. (Adopting a power-oriented approach in the scheduling and register-allocation phases of compilation would provide better chances for subsequent exploitation of the optimized Data Path design here presented, but such modifications to the compiler go beyond the scope of the present work).

Each benchmark program has been compiled with the Lx Compiled Simulator that is part of the Lx toolchain. The Lx compiled simulator accepts a generic C program and processes it in two steps:

1 Translation of the source program into its Lx assembler equivalent;

2 Conversion of the generated Lx assembler program into a new C source that simulates the program run on the Lx architecture (*Compiled Simulator source*). The Compiled Simulator source simulates the Lx architecture by updating a simulated state of the machine each time an assembly statement is executed. Moreover, it collects profiling information to supply a hint on the performance of the program.

For our analysis, the Compiled Simulator source has been instrumented by means of in-house automatic tools so as to trace the following profile information each time a very long instruction is executed, : (i) register definitions; (ii) register uses; (iii) basic block boundaries encountered; (iv) cache misses occurred.

Having compiled and executed the simulator, the generated traces are used to perform the register liveness analysis. To perform such type of analysis, we selected a set of DSP algorithms (written in C) that represent a significant set of target applications for the embedded Lx-architecture. Most of these algorithms were taken from the Mediabench suite [121] while the remaining part of algorithms were in-house optimized versions of classical transforms (see [209, 210] for an overview).

In the case of the two optimized programs (DCT and IDCT), careful algorithm coding leads to lower number of memory accesses and higher register re-use than provided by the original benchmarks. Such tuning grants improved performances.

The average behavior of register liveness for the interval [1-20] is depicted in Figure 8.5, where each point represents the percentage of definitions of a given liveness averaged on all the benchmarks. Note that for liveness lengths greater than 4 the cumulative percentage does not increase significantly.

The behavior of register liveness for the analyzed algorithms are reported in Figure 8.6 where, for a fixed benchmark, each bar represents the percentage of register definitions whose liveness is comprised within a fixed interval. Thus, for example, the black bar represents the percentage of definitions that have a liveness length between 1 and 4 clock cycles.

Even with the simplifying assumptions here made, it can observed that, on average, 28.6% of register definitions has $L_n \in$ [1-2], 35.8% of register definitions has $L_n \in$ [1-3] and that moving to a [1-4] interval does not lead to significant changes with respect to the case $L_n \in$ [1-3]. Note that the peak percentage reaches 71.8% in the case of [1-3] cycles interval which is the interval of interest for our architecture.

In our analysis, we do not take into account the case in which a register is never read between two successive definitions. In fact, register overwriting can occur, for example, across basic blocks or during context switches, but it cannot be statically generated by an optimizing compiler within a basic block. Therefore, this type of register overwriting is out of the scope of our approach, since we are focusing on optimizations applicable within a basic block during the static compilation phase for the VLIW architecture.

6.2 Results of Power Analysis

In order to derive the power savings allowed by our optimization, we performed a transistor-level simulation of the Lx register file, characterized by sixty-four registers and having 8 read-ports and 4 write-ports. We then modeled the register file with an extension of the linear model proposed in [124] that takes into account a base power cost independent from the number of accesses that are performed during a clock cycle, as well as the power costs directly related to read and write operations:

$$P_{RF} = BaseCost + n_w P_{1w} + n_r P_{1r} \tag{8.3}$$

where P_{RF} is the average register file power consumption during a clock cycle, $BaseCost$ is a constant power value due to the activity of the clock signal, n_w is the average number of simultaneous write accesses to the RF (from 0 to 4), P_{1w} is the average power cost of one write access,

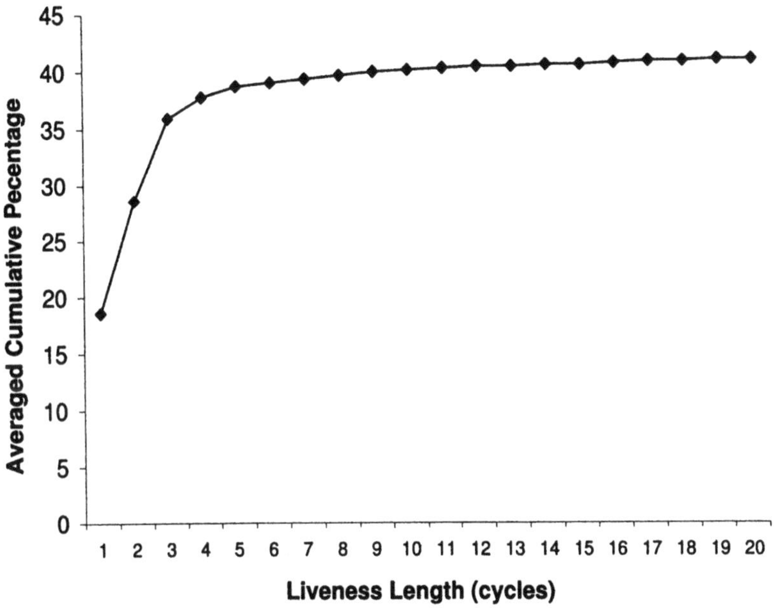

Figure 8.5. Average cumulative distribution of registers liveness

n_r is the average number of simultaneous read accesses to the RF (from 0 to 8), P_{1r} is the average power cost of one read access.

Simulations of the given RF implementation during different and simultaneous RF accesses, led to derive the averaged and normalized power results shown in Figure 8.7. Note that, when the register file is heavily stressed (8 reads and 4 writes per cycle, i.e., the maximum number of allowable simultaneous accesses), the power consumption is 2.5 times the base power cost.

The power analysis of the proposed forwarding optimization has then been derived by combining the power figures of our RF model and the savings in the average number of RF accesses per cycle obtained by profiling each benchmark program through the instrumented Lx Compiled Simulator. Figure 8.8 reports the power saving achieved by applying the proposed low-power solutions to registers whose liveness length is less than or equal to 2 (3) clock cycles. In the case of 2 clock cycles, the average power saving for the given set of benchmarks is 7.4%, while peak saving can reach 25%. In the case of 3 clock cycles, the average power saving for the given set of benchmark is 7.8%, while peak saving can reach 26.1%.

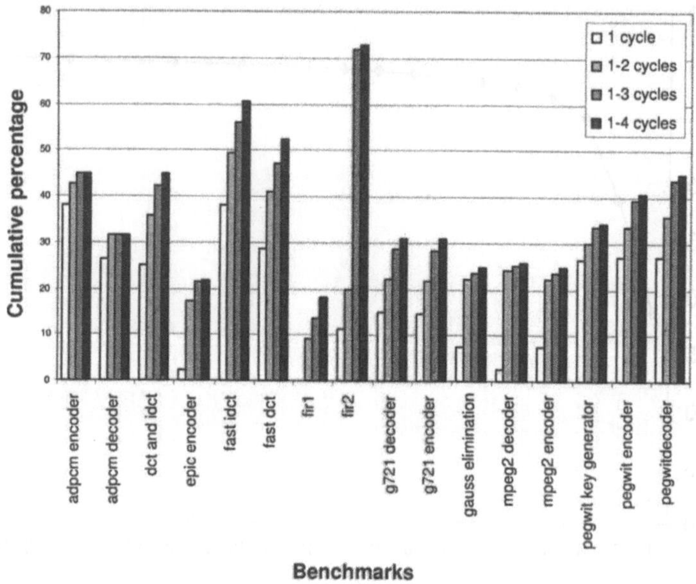

Figure 8.6. Cumulative distribution of registers liveness for the selected benchmark set.

As expected, higher power savings have been obtained by simulating algorithms whose kernel is composed of instructions that tend to maximize short-term reuse of available registers. This approach in fact leads to very short register liveness which can be best exploited by our approach. Instead, the algorithms for which we obtained very low power savings are characterized by a large amount of instruction cache misses and of control operation (such as branches); these tend to limit the application of our optimization even in the presence of a significant amount of short definitions with respect to total definitions. Thus, the particular structure of the code can cause quite different power saving figures even when the register definition behavior of the various algorithms is very similar (e.g. "apcm encoder" and "dct & idct").

7. Extension of the basic idea

The basic idea discussed so far can be extended by considering the set of pipeline micro-registers as constituting a new level in the memory hierarchy, i.e., the microregisters-level. This new memory-level can allow saving space in the RF for variables whose liveness length is very short

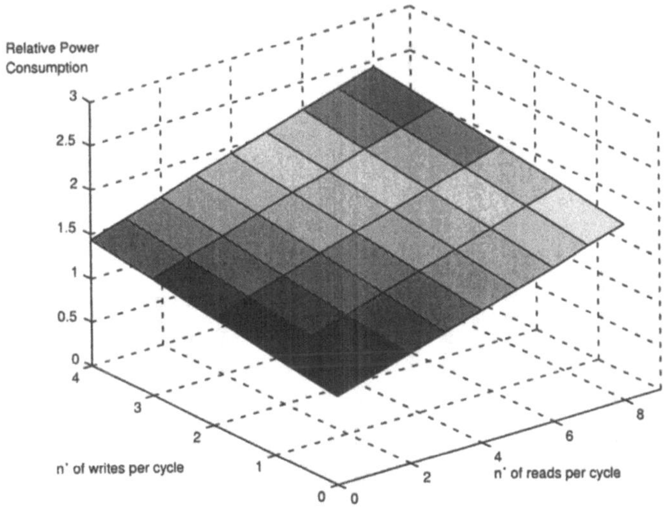

Figure 8.7. Average power consumption of RF read/write accesses (Values are normalized to BaseCost).

and therefore in turn lead to reducing register spilling and ensuing cache traffic.

The compiler, whenever short-lived variables are found such that use of forwarding is applicable and that the conditions specified below are satisfied, does not reserve registers in the RF for such variables. The RF space thus is effectively "increased" as far as compiler use is concerned.

Microregister use becomes a liability when interrupt (and, more in general, exception) handling is concerned; in fact, microregisters can be seen as constituting a "transient" memory, such that could not be associated with a "machine state" to be saved in the case of an exception (unless costly solutions are adopted). By considering interrupts, we see two different possible approaches to overcome this problem:

- *Atomic sequence:* the sequence of instructions using the microregisters is seen as an atomic one, and as such it cannot be interrupted: interrupt is disabled before the starting of the sequence, and the state of the machine is made stable (by writing in the RF or in memory) before the interrupt is re-enabled. This solution does not require any extension of the Instruction Set or of the microarchitecture, and is taken care of by the compiler only.

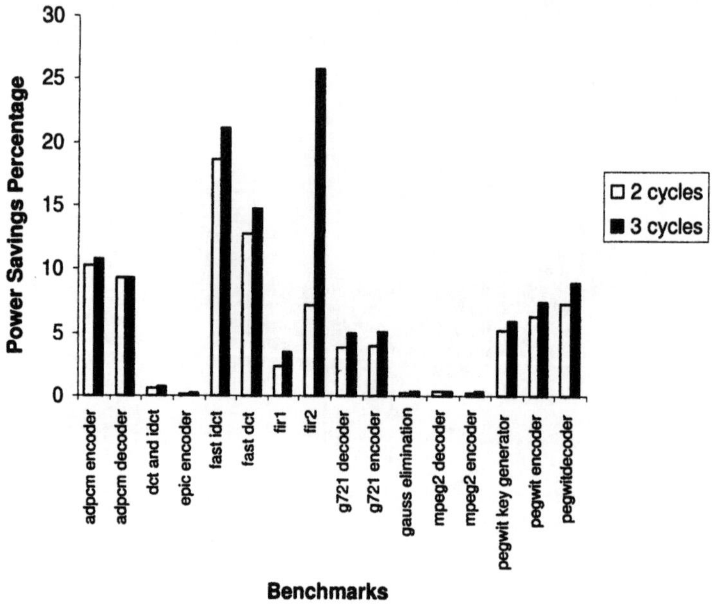

Figure 8.8. Power saving percentage of our optimization approach applied to registers liveness less than or equal to 2 (and 3) clock cycles for the selected set of benchmarks.

- *Checkpointing:* Two new instructions (actually, pseudo-instructions, used by the compiler and affecting only the control unit, not the pipelines) are introduced - namely, checkpoint declaration (`ckp.d`) and checkpoint release (`ckp.r`). At checkpoint declaration, the PC is saved in a shadow register, and until the checkpoint release the machine state cannot be modified (obviously, this implies that no store instructions are allowed); at checkpoint release, the shadow register is reset and interrupts are disabled atomically. Computed results in the checkpointed section can then be committed to the real state of the processor and, after that, interrupts are re-enabled to re-start normal execution. In the case of an interrupt between the `ckp.d` and the `ckp.r`, the PC from which execution will restart after interrupt handling is the one saved in the shadow register (and, obviously, given our constraints on machine state update, the machine state is consistent with that PC). This solution would imply that all register writes in the sequence between ckp.d and ckp.r shall involve microregisters only or that a (small) subset of the RF is reserved for "transient" variables between the checkpoint declaration and release whose lifetime exceeds the maximum one allowed by the pipeline length. Figure 8.9

shows an example segment of code in which check pointing is used; microregisters are indicated with $urNN while normal registers are indicated as usual.

```
                ckp.d                         //save the PC
                sub  $ur18 =  $r9, $r7       ┐
                add  $ur17 =  $r5, $r2       │
                sub  $ur19 =  $r4, $r1;      │
                                             │
                shr  $ur18 =  $ur18, 8       │
                shr  $ur17 =  $ur17, 8       │
                shr  $ur19 =  $ur19, 8;      │
                                             ├ Uninterruptible
                sub  $ur19 =  $ur19, $r8     │    Sequence
                mul  $ur17 =  $ur17, 3784    │
                sub  $ur18 =  $r12, $ur18;   │
                                             │
                sub  $ur19 =  $ur19, $r3     │
                add  $ur18 =  $ur18, $ur17   │
                ckp.r;                       ┘

              ┌ mov $r19=$ur19                // bundle forced
Interrupts    ┤ mov $r18=$ur18                // to WB by ckp.r
 Delayed      └ mov $r17=$ur17;
```

Figure 8.9. Example of a segment of a 4-way VLIW program using checkpointing

8. Conclusions

In this chapter, optimization in view of power consumption has been discussed with reference to the data path of a VLIW microprocessor and to the instructions that access the Register File. The next chapter is devoted to the second power optimization methodology that consists of an efficient exploration of the architectural parameters of the memory sub-systems, from the energy-delay joint perspective. The aim is to find the best configuration of the memory hierarchy without performing an exhaustive analysis of the parameters space. The target system architecture includes the processor, separated instruction and data caches, the main memory, and the system buses.

Chapter 9

A DESIGN SPACE EXPLORATION METHODOLOGY

This chapter presents a more general power optimization technique, aiming at optimizing the system architecture in terms of memory hierarchy with respect to the applications intended for the embedded system. It consists of a system-level design methodology for the efficient exploration of the architectural parameters of the memory sub-systems, from the energy-delay joint perspective. The aim is to find the best configuration of the memory hierarchy without performing the exhaustive analysis of the parameters space. The target system architecture includes the processor, separate instruction and data caches, the main memory, and the system buses. To achieve a fast convergence toward the near-optimal configuration, the proposed methodology adopts an iterative local-search algorithm based on the sensitivity analysis of the cost function with respect to the tuning parameters of the memory sub-system architecture. The exploration strategy is based on the Energy-Delay Product (EDP) metric taking into consideration both performance and energy constraints. The effectiveness of the proposed methodology will be shown through the design space exploration of two real-world case studies: the optimization of the memory hierarchy for an Lx-based system and a MicroSPARC2-based system, both executing the set of Mediabench benchmarks for multimedia applications. Experimental results show an average optimization speedup of 2 orders of magnitude with respect to the full search approach, while achieving either the same optimal solution obtained by the exhaustive system-level exploration or a sub-optimal configuration characterized, in average, by a distance from the optimal full search configuration in the range of 16%.

V. Zaccaria et al. (eds.),
Power Estimation and Optimization Methodologies for VLIW - based Embedded Systems, 143–180.

1. Design space exploration framework

Time-to-market has become the main success factor for many electronic products, in particular those intended to the consumer market. To enable a fast design space exploration while ensuring a good level of optimization in the final architecture, the use of customizable platforms is becoming more and more important [1]. However, even considering simple microprocessor-based architectures composed of a CPU and a memory hierarchy, the search for the optimal system configuration by trading off power and performance still leads to the analysis of very large number of alternatives. Our work aims at overcoming such problems by providing a methodology and a design framework to support the designer in the identification of the optimal solutions in a cost-effective manner. To achieve such a goal, we consider a more abstract view of the system analysis and we identify a suitable set of metrics to carefully predict both the application requirements and their impact on the architecture characteristics. This approach enables a quick analysis of different configurations of the memory sub-system from an energy/delay combined perspective. To support the analysis a flexible design exploration framework has been developed. It is composed of several basic interacting elements, as shown in Figure 9.1,and provides the capability of including different levels in the memory hierarchy, where each level can be characterized by different architectures and implementation technologies.

Three steps characterize the proposed methodology: definition of the system architecture, system characteristics extraction (or *tuning phase*) and design space exploration to determine the optimal system configuration. The next subsections will detail the three steps.

1.1 System description

The first step of the design methodology aims at capturing the relevant characteristics of the embedded application, the customizable implementation platform and the designer's goals.

The design space of the target architecture is defined in terms of the following parameters:

- Number of levels in the memory hierarchy;
- Cache configuration parameters;
- On-chip vs off-chip positioning of cache (L1 and L2);
- Unified vs separate data/instruction cache, at any hierarchy level;
- Width of address and data buses.

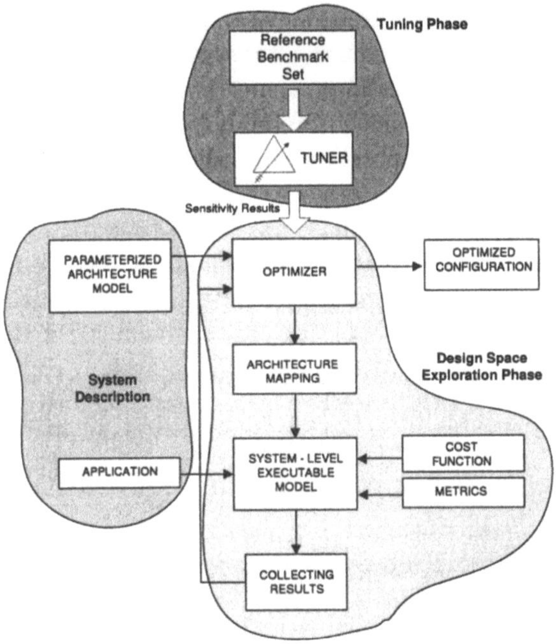

Figure 9.1. The proposed design space exploration flow

- Encoding strategy for the information flowing on both data and address buses.

In the case studies reported in Section 3 and 4, we analyze the instruction and data cache parameters of the target architectures.

During this phase, the designer has to specify the boundaries of the design space for each parameter (for example, the maximum size to be considered for the cache size) and any other constraints on the architectural parameters.

1.2 Design space exploration

To speed up design time while being able to explore different alternative solutions, an efficient system-level trade-off analysis, allowing to early re-target the choice of system parameters, must be defined. The simplest solution would be to perform an exhaustive search of the design space driven by a unique and comprehensive goal function to be optimized. However, this may not be an efficient solution. In fact, let us consider a simple architecture, to be configured according to three parameters, each one characterized by 8 different values: the resulting number of possible configurations to be compared is more than 500 for each of the possible applications of the final system architecture. In such cases, a *brute force* approach could be impractical if the time required by a single simulation is high. Thus, heuristics are needed to reduce the number of configurations to be analyzed, by performing a near-optimal search able to identify acceptable candidate solutions while avoiding long simulation times.

We propose a new heuristic algorithm to explore the design space of microprocessor-based platforms with a configurable memory hierarchy. The heuristic algorithm is implemented in the *optimizer* module of the framework (see Figure 9.1) that is responsible for the generation and the evaluation of new alternatives of the system configuration in order to reach the near-optimal one. The heuristic algorithm is driven by the characteristics of the system extracted by means of a sensitivity analysis performed during the tuning phase. Hence, the heuristic algorithm is called the *sensitivity-based search* algorithm.

The goal of the heuristic algorithm is to move towards a restricted set of candidate optimal configurations, by avoiding the exhaustive analysis of the design space. This heuristic method is tuned once for each specific platform by analyzing its behavior while executing a representative set of benchmarks. The goal is to identify the impact of the different parameters on the cost function to be optimized by performing a comparative analysis of the magnitude of the discrete partial derivatives of the cost function. This information is used to efficiently drive the strategy of the sensitivity-based search engine, that will be used to optimize each new application for that specific platform.

In practice, instead of considering the simultaneous optimization of n parameters, requiring $O(k^n)$ simulations (where k is the average range of variability of each parameter), the heuristic explores sequentially the region of variation of the most important parameters, thus producing near-optimal solutions (in terms of minimization of the cost function). In fact, the analysis considers one parameter at a time in order of decreasing sensitivity, based on the results of the tuning phase for the given platform. The number of simulations, in the average case, is linear with respect to the range of variations of each parameter and, as shown in the experimental results (see Section 4), the optimal solution is almost always identified thanks to the accuracy of the sensitivity analysis performed during the tuning phase.

1.3 Computing the cost of a configuration

The *cost function* of a system provides a quantitative measure of the system characteristics with respect to the parameters composing the function, and allows the system optimization by discarding non-optimal solutions while considering several metrics simultaneously. In this work, we consider power and performance as the most relevant metrics for our target embedded systems, with the aim of minimizing power consumption while maximizing performance. These metrics have been extensively discussed in the literature [150, 151] and we adopted the Energy-Delay Product (EDP) as cost function to compare alternative system designs.

To compute such a general goal function for the system platform, we focus on the energy and the delay associated with the processor and memory parts of a system, considering the contributions of the processor core, the system-level buses and each level of the memory hierarchy. While the computation of delays is part of the cycle-accurate simulation environment (the *System Executable Model*, see Figure 9.1), the evaluation of the overall energy requires the extraction of the actual use of each system resource during the execution of the application. These data are necessary to evaluate the analytical energy models defined for each system component that compose the overall energy metric. All these statistics are gathered by profiling the functional and memory behavior of the processor during the simulation of the embedded application by means of a cycle-accurate Instruction-Set Simulator (ISS).

The analytical energy models include the following contributions:

- processor core;
- processor I/O pads;
- processor-to-$L1$ on-chip buses;
- on-chip $L1$ I- and D-caches;
- $L1$-to-$L2$ off-chip buses;
- $L2$ unified $SRAM$ cache;
- Main memory ($DRAM$).
- $L2$-to-$DRAM$ off-chip buses;

Energy and delay values are normalized with respect to the instructions executed in order to allow the comparison of the behavior of different applications on the same system architecture or of different architectures on the same application.

Let us overview the energy models of each system component. For each processor core, a suitable model or tool able to provide the power consumption at the instruction-level is needed [99, 101, 112]. In this version of the design environment, we used a simple instruction-level model that accounts for the average power consumed by the processor in normal operating conditions and we multiplied it by the average execution time of an instruction.

For the system-level buses (such as the processor-to-$L1$ buses), the energy model can be simply expressed as: $E \approx (C_{load} V_{dd}^2 n_{trans})/2$, where C_{load} is the bus load capacitance, V_{dd} is the power supply voltage and n_{trans} is the number of transitions on the bus lines, depending on the hamming distance between subsequent vectors composing the bus stream.

For on-chip $L1$ caches, we use the analytical energy model developed in [125], that accounts for:

- Technological parameters (such as capacitances and power supplies);
- Architectural parameters (such as block size and cache size);
- Switching activity parameters (such as number of bit line transitions).

This cache energy model has been used in recent works (such as in [150]), where the switching activity parameters have been computed either by using application-dependent statistics or by assuming typical values (such as half of the address lines switching during each memory request). In our framework, we directly import the actual values of hit/miss rates and bit transitions on cache components, that have been derived by the system-level simulation environment to account for the actual profiling information depending on the execution of embedded applications.

Finally, the main memory power characterization has been simply obtained by directly importing the information derived from the datasheet into our model.

2. Sensitivity-Based Optimization

The optimization methodology is based on the *sensitivity* analysis of the system with respect to a given set of configuration parameters. Before detailing the sensitivity based optimization, let us introduce the notions of architectural space and the sensitivity associated with the cost function.

2.1 Architectural space definition

The architectural optimization algorithm is a local search algorithm whose target is to find a configuration (or *point*) in the *architectural space* whose cost is (locally) minimum.

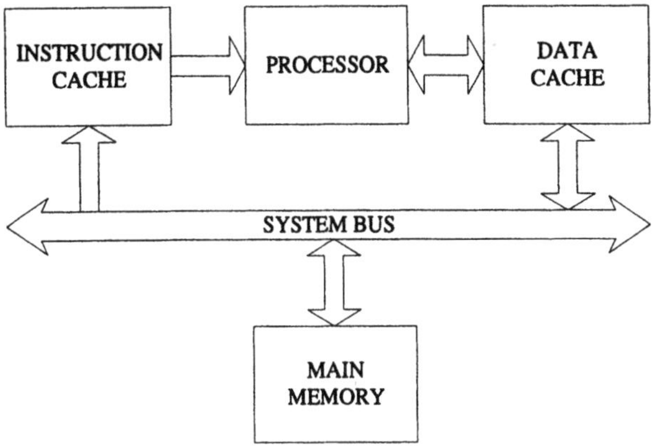

Figure 9.2. The target system architecture.

Given a set $P = \{p_1, \ldots, p_n\}$ of n architectural parameters, the architectural space $\mathcal{A}$ can be defined as:

$$\mathcal{A} = S_{p_1} \times \ldots S_{p_l} \ldots \times S_{p_n} \tag{9.1}$$

where S_{p_l} is the ordered set of possible configurations for parameter p_l and "$\times$" is the cartesian product. As an example of architectural space, let us consider the configurable system architecture shown in Figure 9.2 composed of a processor, separate I- and D- L1 caches, the main memory and the system buses. Let us consider the exploration of the design space of this architecture in terms of six parameters: cache size, block size and associativity of both D- and I- L1 caches. In this case, each instance of the configurable architecture is described as a *6-tuple* $[c_i, b_i, v_i, c_d, b_d, v_d] \in \mathcal{A} = S_{c_i} \times S_{b_i} \times S_{v_i} \times S_{c_d} \times S_{b_d} \times S_{v_d}$ where:

- S_{c_i}, S_{c_d} are the ordered sets of the possible sizes of the I- and D-caches (e.g. {16KB, ..., 256KB}).

- S_{b_i}, S_{b_d} are the ordered sets of the possible block sizes of the I- and D-caches (e.g. {16B, ..., 64B})

- S_{v_i}, S_{v_d} are the ordered sets of the possible associativity values of the I- and D-caches (e.g. from direct mapped to 8-way set associative).

Assuming that each set S of possible configurations is discrete and ordered ($S = \{s_1, s_2, \ldots s_o \ldots, s_q\}$), we can define a *neighborhood* operator "$\rightarrow$":

$$s_o \rightarrow n = \begin{cases} s_{o+n} & 1 \leq o + n \leq q \\ \perp & \text{otherwise} \end{cases} \tag{9.2}$$

where "$\perp$" stands for "indefinite value". As an example of application of the neighborhood operator, if we consider the following I-Cache size space:

$$S_{c_i} = \{16KB, 32KB, 64KB, 128KB, 256KB\} \tag{9.3}$$

we have that $(64KB \rightarrow 1) = 128KB$ while $[16KB \rightarrow (-1)] = \perp$.

The neighborhood operator defines positive and negative *perturbations* of a configuration. Given an architecture

$$a = [a_{p_1} \ldots a_{p_l} \ldots a_{p_n}] \in \mathcal{A} \tag{9.4}$$

and a parameter p, we define a perturbation $\delta_p(a, y)$ as a vector

$$b = [b_{p_1} \ldots b_{p_l} \ldots b_{p_n}] \in \mathcal{A} \tag{9.5}$$

where

$$b_{p_l} = \begin{cases} a_{p_l} & p_l \neq p \\ a_{p_l} \rightarrow y & p_l = p \end{cases} \tag{9.6}$$

Let us denote by $\delta_p^+(a)$ the perturbation $\delta_p(a, 1)$ and by $\delta_p^-(a)$ the perturbation $\delta_p(a, -1)$. In other words, a perturbation is an operation that modifies only a single element of the configuration at a time (i.e., the element corresponding to p), by applying the neighborhood operator.

The next section specifies the problem of finding the architectural optimal configuration with respect to a defined cost function and presents its solution through the sensitivity-based optimization algorithm.

2.2 Sensitivity of the cost function

The proposed methodology aims at identifying an optimal configuration of a given architecture platform for performance and energy consumption [211, 212]. To assess the optimization problem, we need to introduce a cost function associated with the architecture under consideration and with the applications to be executed. A cost function is defined as a function $f_M : \mathcal{A} \times \mathcal{K} \rightarrow \mathcal{R}$, where $\mathcal{K}$ is the set of applications that can be executed on the architecture, $\mathcal{R}$ is the set of real numbers representing the cost of the given system configuration and M is the machine model under consideration. The machine model is used to differentiate among the cost functions associated with different families of architectures (e.g. M="SunSPARC" or "x86").

The cost function adopted in this work is defined as the product of two parameters, i.e., energy and delay:

$$f_M(a, k) = \text{Energy}_M(a, k) \times \text{Delay}_M(a, k) \tag{9.7}$$

where $\text{Energy}_M(a, k)$ is the energy dissipated by the system in configuration a during the execution of application k and $\text{Delay}_M(a, k)$ is the number of cycles necessary to the system in configuration a to execute k. Since both energy and delay associated with the overall system can be hardly expressed by analytical functions, we adopt an executable model (i.e., a simulation-based model) to compute the values of the energy and delay parameters associated with each point of the parameter space.

The basic idea of our approach is to exploit the influence of the architectural parameters on the cost function to minimize the simulations to be carried out during the optimization phase. We introduce the concept of *sensitivity* to express the variability of the cost function in the neighborhood of the minimum for the given set of reference benchmark applications. The reference benchmarks have been chosen in order to stress the functionality of the architecture in terms of memory accesses, use of functional units and variability of all the relevant parameters.

The sensitivity of the cost function with respect to a parameter p and a set of reference applications $\mathcal{H}$ is defined as:

$$\sigma(p, f_M) = E_{h \in \mathcal{H}} \left[\frac{\Delta_p(f_M(a_{opt}(h), h))}{f_M(a_{opt}(h), h)} \right] \tag{9.8}$$

where E is the mean operator over the set of reference applications $\mathcal{H}$, $a_{opt}(h)$ is the full search optimal configuration found for h, M is the machine model and $\Delta_p(f(a,h))$, is defined as:

$$\Delta_p(f(a,h)) = |f(a,h) - max(f(\delta_p^+(a), h), f(\delta_p^-(a), h))|; \tag{9.9}$$

Equation 9.8 defines the average percentage variation of the cost function with respect to a small perturbation from its minimum. More specifically, the variation $\Delta_p(f(a,h))$ is the maximum absolute variation due to a small positive or negative perturbation on the optimum of the cost function for a given application h.

2.3 Sensitivity-based local search algorithm

To efficiently identify an optimal (or sub-optimal) configuration that minimizes the cost function our approach takes into account not only the architectural parameters and their influence on the cost function, but also the target applications that will be executed on the embedded system.

Given the architectural space $\mathcal{A}$, the application k and the cost function f_M, the optimization problem can be stated as

$$\text{find } a_{opt} \text{ such that } f(a_{opt}, k) = \min_{\forall a \in \mathcal{A}} f_M(a, k). \tag{9.10}$$

To solve the problem expressed by Equation 9.10, we define a two-steps methodology. The first step consists of a tuning phase that analyzes the architecture platform in terms of *sensitivity*. The second step

```
TuningPhase(f_M, H, P)
{
    Input:   f_M=cost function derived from machine model
             H=set of reference applications to be optimized;
             P=set ofconfiguration parameters;
    Output:  σ= set of sensitivity values for each parameter;

    foreach h ∈ H
    {
        a_opt(h)=find full search optimum of h, given f_M;
        foreach p ∈ P
        {
            σ(p, f_M) = σ(p, f_M)+[Δ_p(f_M(a_opt(h),h)) / f_M(a_opt(h),h)] × 1/|H|
        }
    }
}
```

Figure 9.3. The algorithm of the tuning phase.

consists of the actual local search algorithm, that exploits the sensitivity information determined in the tuning phase, to perform the efficient exploration of the architecture with respect to the target applications. In fact, the sensitivity information drives the design space exploration, allowing the identification of the path that provides the largest improvement on the cost function, and thus a faster convergence to the minimum.

The tuning phase is applied only once for a particular set of reference applications $\mathcal{H}$ and a machine model M, while the local-search algorithm is applied to every new application to be optimized on the system.

The basic steps of the tuning phase are shown in Figure 9.3. Given a set $\mathcal{H}$ of reference applications, the tuning phase consists of a full architectural space exploration for each $h \in \mathcal{H}$, by computing the finite difference Δ_p and finally the average sensitivity for each parameter p of the architecture. A set of reference applications has to be selected to reflect the typical behavior of the given architecture in terms of accesses to the memory hierarchy and to the function units.

Once the sensitivity $\sigma(p, f_M)$ has been characterized, the sensitivity-based local search algorithm reported in Figure 9.4 can be applied to any new application for which the optimal architecture must be determined. The algorithm receives as inputs:

- the cost function f_M and its sensitivity information σ;
- the application k to be optimized;
- the starting point for the exploration phase a_0.

```
SensitivityBasedLocalSearch(f_M,k,σ,P,a_0)
{
    Input:   f_M=cost function derived from machine model
             k=application to be optimized;
             σ=vector of sensitivity values;
             P=set of configurable parameters;
    Output:  a_opt= optimal configuration for application k;

    L= order P by sensitivity σ;
    a_opt = a_0;
    do
    {
        foreach p ∈ L  /*visit the parameters in order of sensitivity*/
        {
            find n such that f_M(δ_p(a_opt,n),k) ≤ f_M(δ_p(a_opt,y),k),∀y;
            a_opt = δ_p(a_opt,n);
        }
    } while improvements on f_M are negligible
}
```

Figure 9.4. The sensitivity-based local search algorithm.

The starting point is chosen as the average of the optimal architectures identified during the tuning phase:

$$a_0 = E_{h \in \mathcal{H}} \left[a_{opt}(h) \right] \tag{9.11}$$

The sensitivity-based algorithm selects iteratively one parameter at a time and attempts to decrease the sensitivity and to optimize the cost function with respect to the chosen parameter, by varying its value, through acceptable perturbations. The stopping criterion is defined as the point in which the cost function improvement is below a given threshold, thus indicating that further attempts of optimization in that direction will not produce sizeable advantages in the cost function reduction.

To clarify the proposed approach, let us consider a simple example consisting of an application k running on a system characterized by three configurable parameters (cache size c_i, cache block size b_i and cache associativity v_i). Let us assume that the sensitivity analysis has shown that the cost function f_M is affected first by c_i, second by v_i, and third by b_i. Given a suitable a_0 (see Equation 9.11), the optimization algorithm executes the following steps:

1 $a_{opt} = a_0$

2 $oldcost = f_M(a_{opt}, n)$;

3 fix the component b_i and v_i of a_{opt} and find the optimal c_i such that $f_M(a_{opt}, k)$ is minimized;

4 fix the component c_i and b_i of a_{opt} and find the optimal v_i such that $f_M(a_{opt}, k)$ is minimized;

5 fix the component c_i and v_i of a_{opt} and find the optimal b_i such that $f_M(a_{opt}, k)$ is minimized; $newcost = f_M(a_{opt}, n)$.

6 if $(\frac{newcost-oldcost}{oldcost} > threshold)$ go back to step 2.

where *threshold* is a fixed value that defines the stopping criterion for the search of the local optimum.

Therefore, for each new application, the result of the sensitivity analysis of the system will be preserved, so that instead of considering the simultaneous optimization of p parameters only a narrower region of possible configurations producing sub-optimal solutions will be explored.

The sensitivity based optimization requires a number of simulations equal to:

$$g \times \sum_{p_l \in P} |S_{p_l}| \tag{9.12}$$

where P is the set of configurable parameters, S_{p_l} is the ordered set of configurations associated with p_l and g is the number of iterations over the whole set of parameters of the algorithm.

Therefore, the speed-up of the proposed approach,with respect to a full-search algorithm, can be computed as:

$$\text{Speedup} = \frac{|\mathcal{A}|}{g \times \sum_{p_l \in P} |S_{p_l}|} = \frac{\prod_{p_l \in P} |S_{p_l}|}{g \times \sum_{p_l \in P} |S_{p_l}|} \tag{9.13}$$

The obtained speed-up represents a relevant improvement in terms of simulation time if the number g of iterations is small. In fact, the choice of the starting point a_0 as defined in Equation 9.11, increases the probability of a small number of iterations g, since there is a high probability that the optimum solution is in the neighborhood of a_0.

3. The Lx architecture: a case study

In this section, we show how the Lx architecture can be efficiently optimized by the application of the sensitivity-based local search algorithm. The target system is composed of the 200 MHz Lx1 processor core (operating at 3.3V), separated and configurable on-chip I- and D-caches with one-cycle hit time, a processor-to-memory bus at 200MHz and 8MByte of DRAM as main memory.

Each instance of the virtual architecture has been described as a *6-tuple* $[c_i, b_i, v_i, c_d, b_d, v_d] \in \mathcal{A} = S_{c_i} \times S_{b_i} \times S_{v_i} \times S_{c_d} \times S_{b_d} \times S_{v_d}$ where:

- S_{c_i}, S_{c_d} are the ordered sets of the possible sizes of the I- and D-caches ({2KB, 4KB, 8KB,16KB, 32KB, 64KB}).
- S_{b_i}, S_{b_d} are the ordered sets of the possible block sizes of the I- and D-caches ({8B, 16B, 32B})
- S_{v_i}, S_{v_d} are the ordered sets of the possible associativity values of the I- and D-caches ({1, 2, 4, 8})

The architecture has been explored by using the power oriented Lx-ISS presented in Chapter 6, in which the cache energy models (tailored to the specific configuration of Lx) have been substituted with the CACTI [122] configurable cache energy and delay model. This configurable ISS can be used to gather the following statistics:

- Number of accesses to the memory hierarchy (on both I- and D-buses);
- Average cache miss rate (on both I- and D-buses);
- Address sequentiality (on both I and D address buses) [160];
- Bus transition activities (on both I and D address buses);
- Delay D (average clock cycles per instruction, measured in terms of $[CPI]$);
- Energy E dissipated by the architecture during program execution measured in terms of *Joule per Instruction* $[JPI]$.

The benchmarks used to validate the model are part of the Mediabench suite [150] (see Appendix 2), a set of multimedia benchmarks that can be considered a *de-facto* standard for multimedia applications. The benchmarks have been executed (with the standard options provided in their makefile) by the customized Lx-ISS to provide system statistics.

To give an idea of the energy-delay trend for the target system, Figure 9.5 reports the scatter plot of the energy and the delay for each point in the architectural space (actually composed of 5184 points) for the `jpegenc` benchmark. In this picture, we can observe the system with the absolute minimum delay (point A), with the minimum energy (point B), with the minimum EDP value (point C) and with the current Lx configuration (see Chapter 6). The EDP optimal point is, with respect to the most performant configuration A, 4% slower but saves as much as 240% in energy consumption while, with respect to the less energy consuming configuration B, C is 19% faster and it is only 2.9% more energy consuming. Note that, going from A to C (B to C), the percentage of energy (performance) savings are always greater than performance (energy) losses. For what concerns the current Lx configuration (computed with the CACTI cache model), it can be seen that it consumes 40% power more than the optimal EDP configuration, presenting almost the same performance.

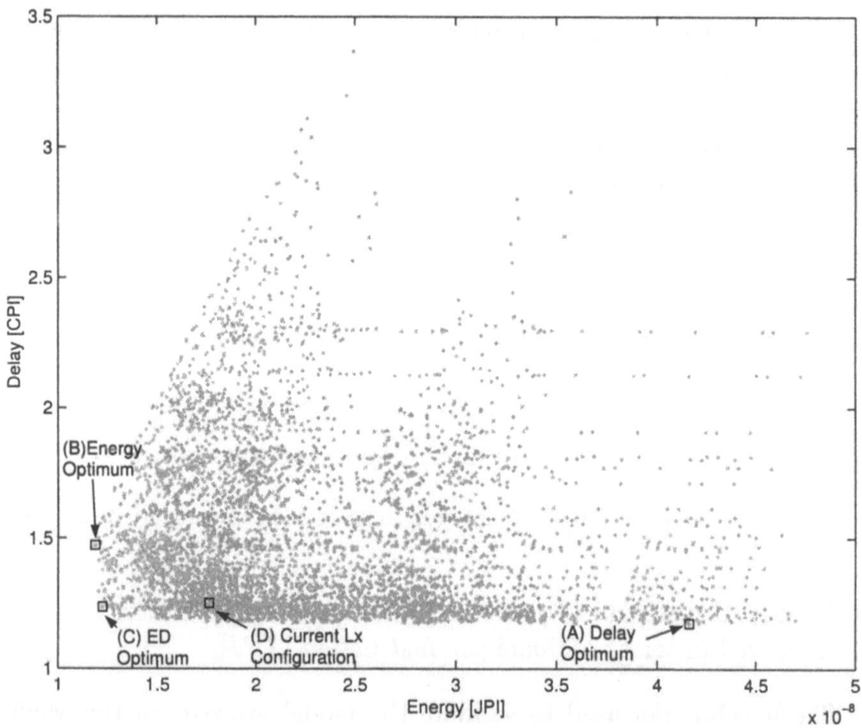

Figure 9.5. Normalized delay $[CPI]$ vs. energy $[JPI]$ scatter plot for the `jpegenc` benchmark.

Let us analyze more in detail the benchmark `jpegenc` in order to better understand the behavior of the system in terms of delay and energy metrics. Figures 9.6 and 9.7 show the delay and the energy-delay product in the neighborhood of the optimal configuration of the system for the benchmark `jpegenc` ($a_{opt} = [16384, 32, 1, 32768, 32, 1]$) by varying the cache sizes and block sizes.

Note that, the increase of the cache size, while keeping a constant block size, tends to minimize the delay since more blocks are present in the cache and the miss probability decreases. However, increasing the cache increases also the energy consumption and this implies a trade-off between energy consumption and performance. This is clearly shown in Figure 9.7, where the energy-delay curves for the I-cache are significantly convex, while for the D-cache the effect is reduced.

Figure 9.8 and 9.9 show the delay and the energy-delay product in the neighborhood of the optimal configuration of the system for the benchmark `jpegenc` by varying the cache sizes and set associativity of both caches. Increasing the cache size with a constant associativity increases the energy consumption. The increased associativity leads to reduced delay but impacts substantially on the overall energy-delay product. Also in this case the I-cache size must be balanced to find an optimal energy and delay congiguration.

Figure 9.10 and 9.11 show the delay and the energy-delay product in the neighborhood of the optimal configuration of the system for the benchmark `jpegenc` by varying the block sizes and associativity of both caches. We can observe how increasing the block size with a constant associativity decreases the delay, even if the data transfer time from the main memory becomes longer. This is due to the fact that the cache miss ratio is reduced significantly.

However, increasing the associativity implies a substantial energy overhead that impacts on the energy delay product. This is due to the CACTI energy model, that assigns a very high weight to the additive control logic required to manage the associativity. This forces us to use very simple caches (direct mapped), to minimize the energy delay product.

3.1 Application of the Tuning Phase

The tuning phase is based on the execution of the MediaBench suite (see Appendix 2). Among the Mediabench, we selected a sub-set (shown in figure 9.1) of control-intensive and data-intensive applications to be used as reference application set $\mathcal{H}$ to characterize the sensitivity of the system.

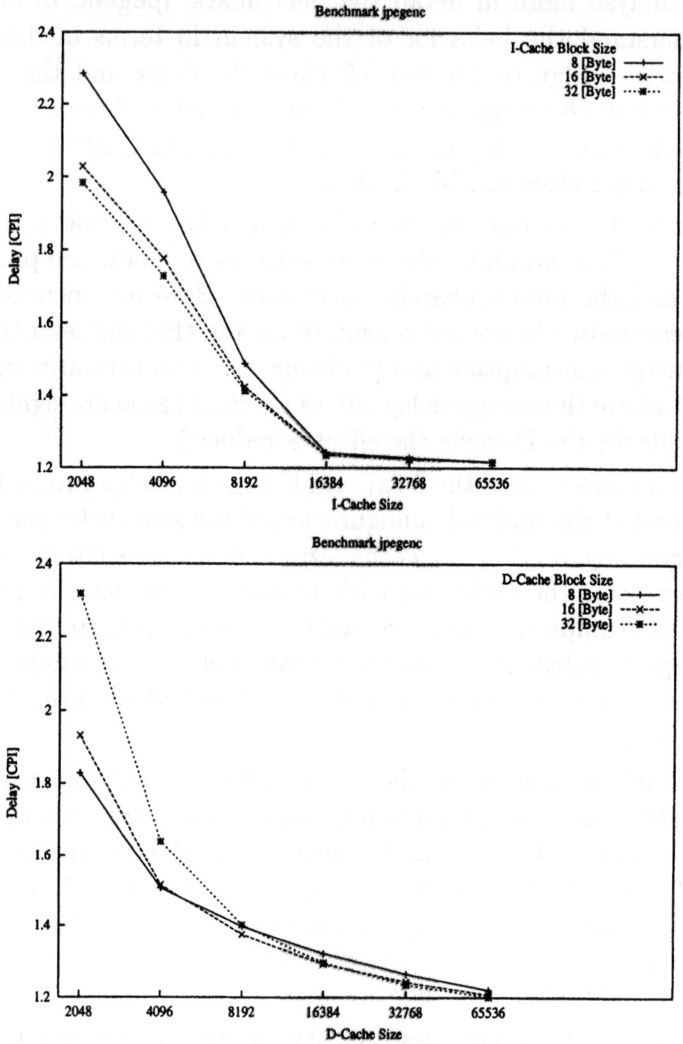

Figure 9.6. Delay for *jpegenc* benchmark, by varying cache sizes and block sizes

Table 9.2 shows the optimal parameter values found for the reference benchmarks and their sensitivity indexes. The last row (AVG) indicates the average sensitivity value associated with the parameter as well as the feasible value of the parameter closest to the average of the optimum.

If we consider the average sensitivity, as Figure 9.12 shows, the most important parameters are in order, the I-cache size (c_i), the I-cache associativity (v_i), the I-cache block size (b_i), the D-cache associativity (v_d), the D-cache size (c_d) and D-cache block size (b_d). In fact, as figures 9.7-9.11 show, the I-cache size and I-cache associativity have a strong

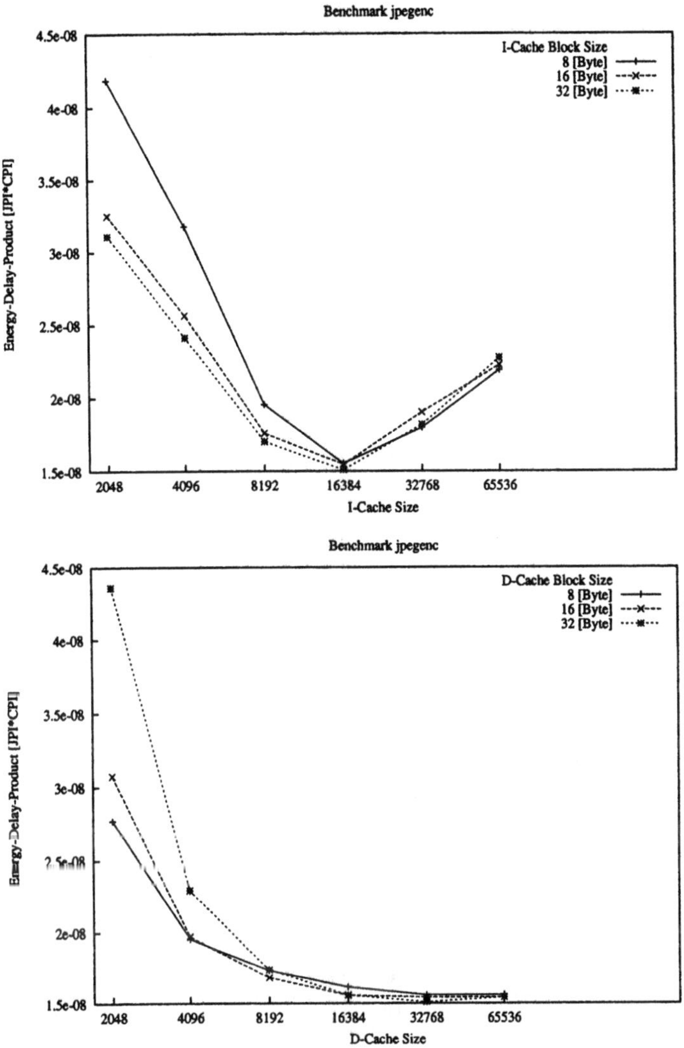

Figure 9.7. ED product for *jpegenc* benchmark, by varying cache sizes and block sizes

impact on the Energy Delay Product cost function. Thus, by using as a starting point the average configuration shown in Table 9.2, the sensitivity algorithm tries to improve the speed of convergence by optimizing first these two parameters and then the remaining ones, as shown in the next section.

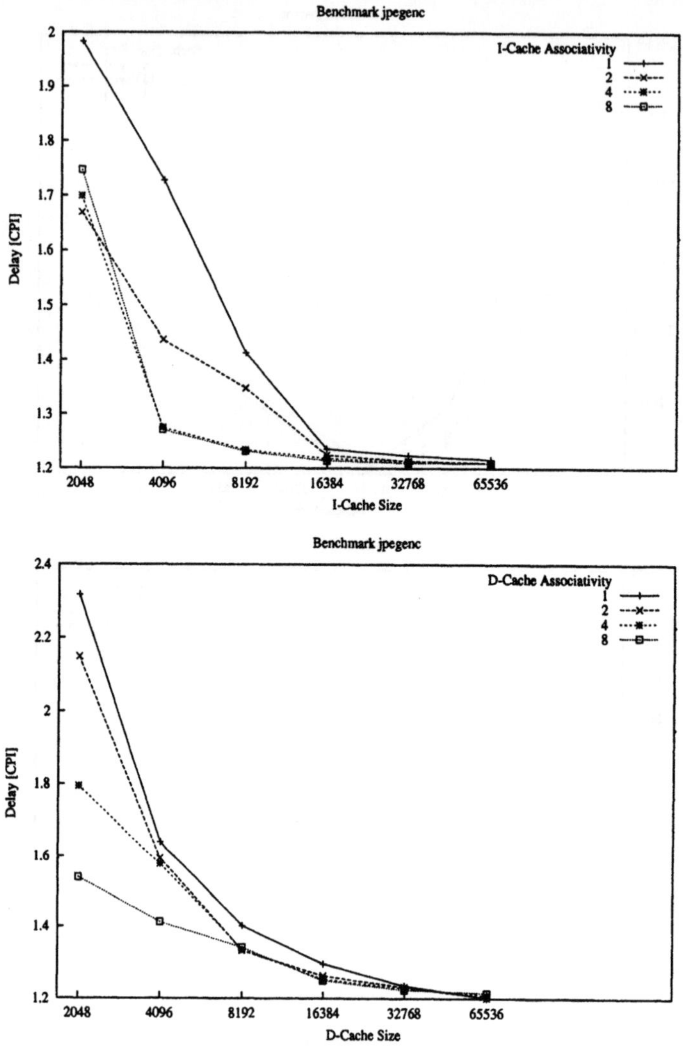

Figure 9.8. Delay for *jpegenc* benchmark, by varying cache sizes and associativity

3.2 Sensitivity-Based Optimization

To validate the methodology, we applied the sensitivity based optimization to a different sub-set of Mediabench applications. The sub-set is disjoint from the set of reference benchmarks used during the tuning phase while the sensitivity values used for this optimization have been presented in the previous section. For validation purposes, the applications shown in table 9.3 have been used.

Figure 9.13 provides the behavior of the sensitivity-based optimizer in the energy-delay space. As shown, the algorithm starts from a point

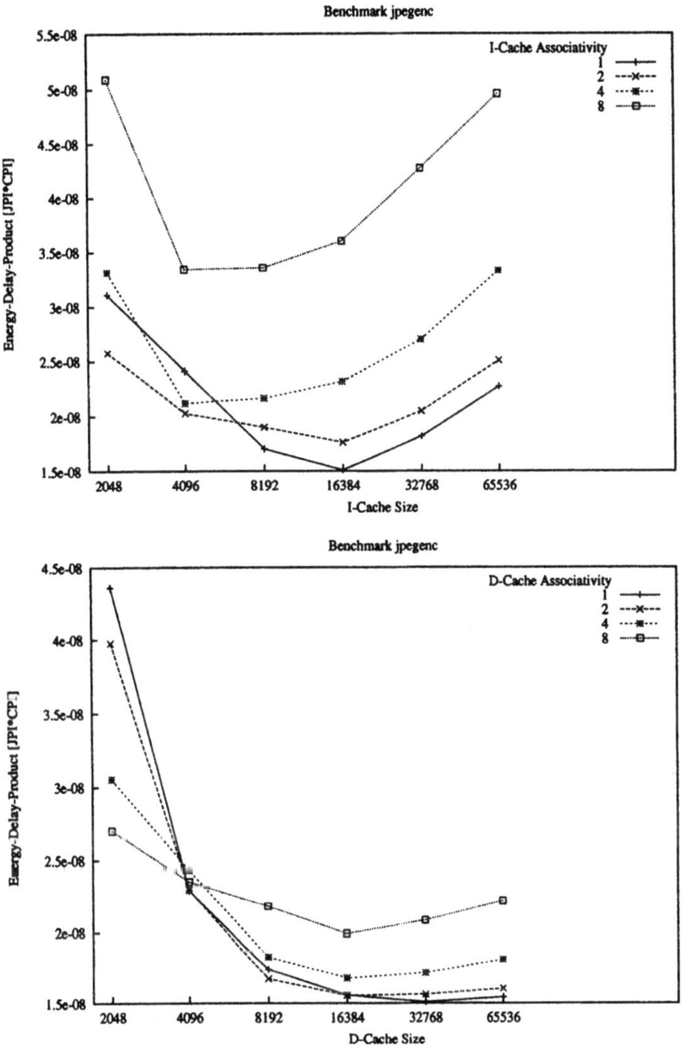

Figure 9.9. ED product for *jpegenc* benchmark, by varying cache sizes and associativity

characterized by high energy and delay values and tends to optimize energy and delay simultaneously. As a matter of fact, it reaches first a best delay configuration and, from this point, it begins to consider all the points of the pareto curve until it arrives to the best energy-delay configuration.

Table 9.4 shows the best configurations estimated by the sensitivity optimizer with respect to those found with a full search over the entire design space. The table shows also the percentage errors on the energy-

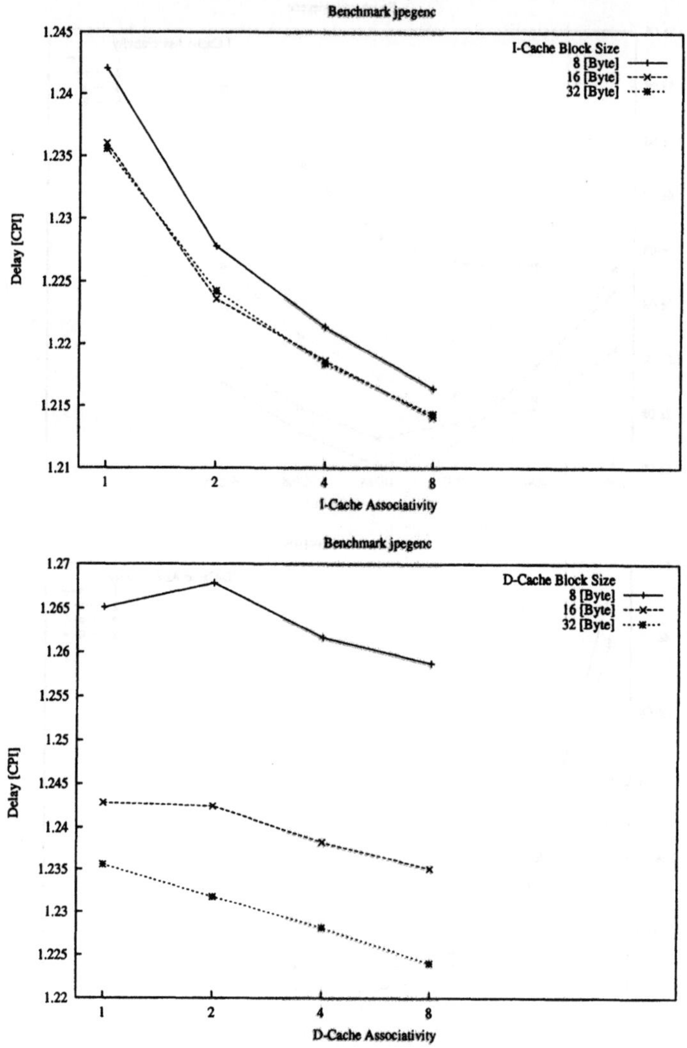

Figure 9.10. Delay for *jpegenc* benchmark, by varying block sizes and associativity

delay product between the real and the estimated optimum. Note that for one benchmark, the configuration found by the heuristic corresponds to the full search one and that the average error is less than 16%. For what concerns the **g721dec**, the overall EDP error is almost 74%, almost entirely due to an energy inefficient, but very performant, local minimum.

Figure 9.14 reports the speedup in terms of number of simulations needed by the sensitivity optimizer. As the figure shows, the average ratio between the number of simulations performed by the full search

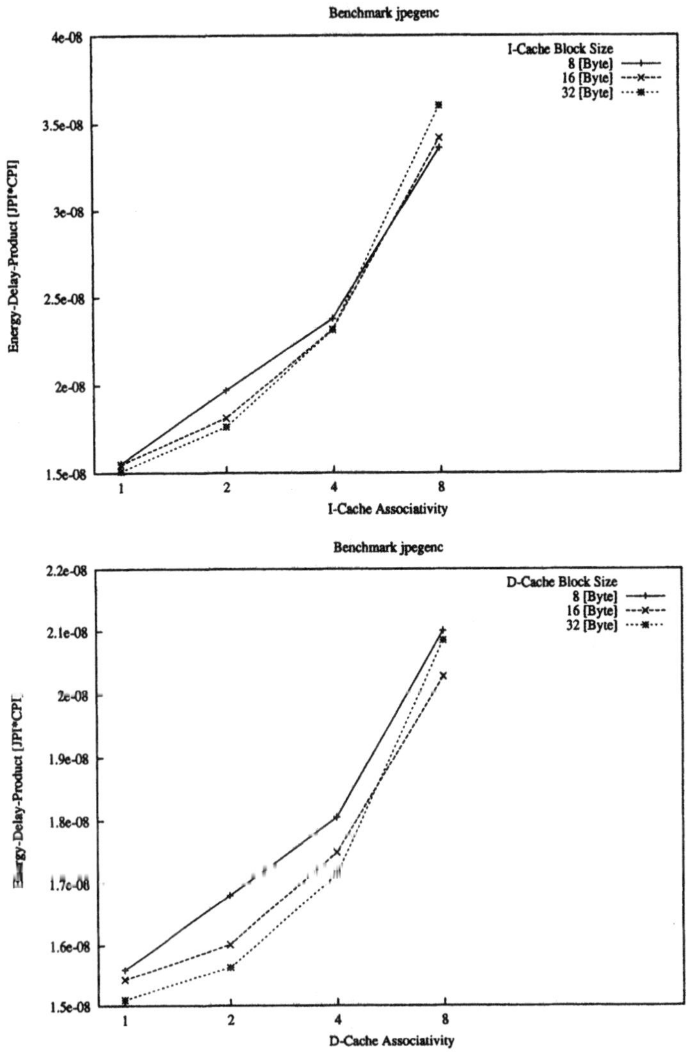

Figure 9.11. ED product for *jpegenc* benchmark, by varying block sizes and associativity

and the number of simulations performed by the sensitivity optimizer is, on average, two orders of magnitude larger. As an example, for a generic application requiring a simulation time of two minutes for each single simulation step, the full search optimization would have required 13 days approximately, while our methodology finds a near optimum solution in less than 4 hours.

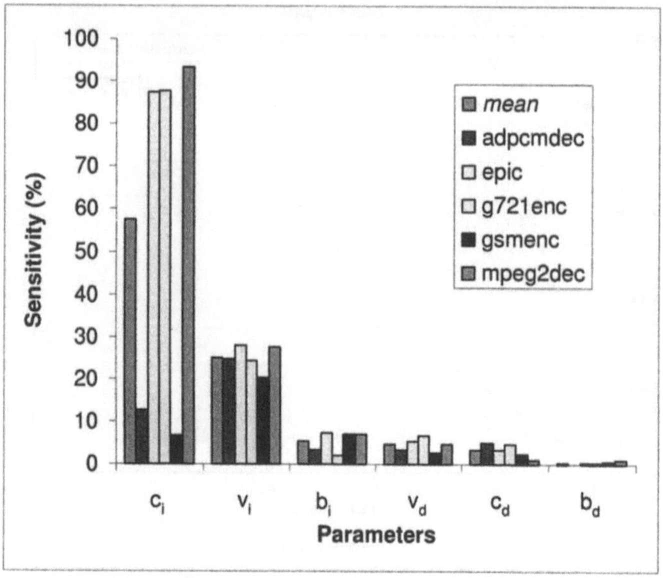

Figure 9.12. The distribution of the sensitivity for the set of the architectural parameters ordered by their mean sensitivity

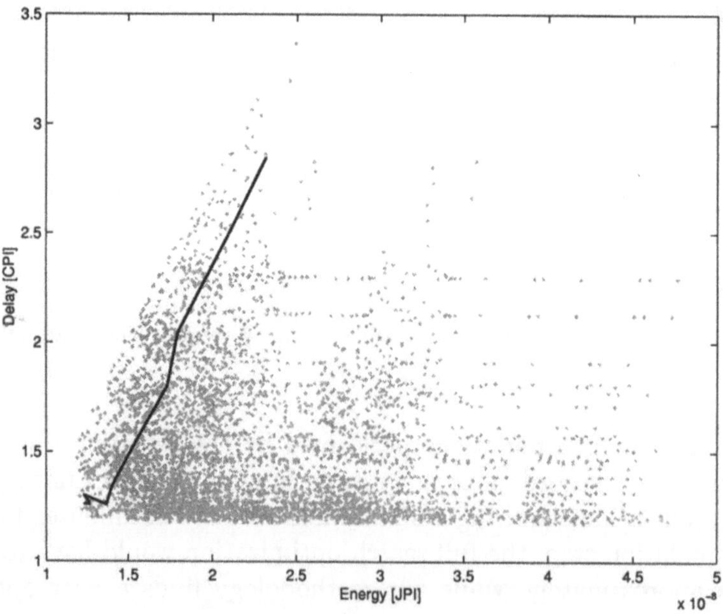

Figure 9.13. The path followed by the sensitivity-based algorithm when optimizing the **jpeg** encoder application on the **Lx** system

Benchmark name	Abbreviation
Adaptive Differential Pulse Code Modulation Decoder	adpcmdec
Wavelet compression algorithm	epic
G721 voice compression encoder	g721enc
GSM encoder	gsmenc
MPEG2 Digital image stream compression decoder	mpeg2dec
PEGWIT Cryptographic algorithm	pegwit

Table 9.1. The set of benchmarks used for the sensitivity analysis.

Bench.	c_i	$\sigma(c_i)$	b_i	$\sigma(b_i)$	v_i	$\sigma(v_i)$	c_d	$\sigma(c_d)$	b_d	$\sigma(b_d)$	v_d	$\sigma(v_d)$
adpcmdec	$8K$	12.5%	32	3.2%	1	24.6%	$16K$	5.1%	8	0.1%	1	3.2%
epic	$32K$	87.5%	8	7.1%	1	27.7%	$8K$	3.3%	16	0.4%	1	5.3%
g721enc	$16K$	87.6%	32	1.9%	1	24.1%	$4K$	4.6%	16	0.4%	2	6.6%
gsmenc	$16K$	6.7%	32	7.1%	1	20.3%	$4K$	2.2%	16	0.5%	1	2.7%
mpeg2dec	$32K$	93.4%	8	6.9%	1	27.5%	$2K$	0.9%	8	1.0%	1	4.6%
pegwit	$8K$	15.3%	16	3.4%	1	20.6%	$64K$	20.4%	8	15.6%	1	11.0%
AVG	$16K$	57.5%	16	4%	1	24.8%	$8K$	3.2%	16	0.5%	1	4.5%

Table 9.2. The set of optimal parameters and sensitivity indexes for the chosen set of reference applications.

Benchmark name	Abbreviation
Adaptive Differential Pulse Code Modulation Encoder	adpcmenc
G721 voice compression decoder	g721dec
GSM decoder	gsmdec
Jpeg image compression encoder	jpegenc
Jpeg image compression decoder	jpegdec
MPEG2 Digital image stream compression encoder	mpeg2dec

Table 9.3. The set of benchmarks used for the validation of the sensitivity approach

Bench.	a_{opt}	$ED(a_{opt})$	$\hat{a}_{opt}$	$ED(\hat{a}_{opt})$	*error*
adpcmenc	$[8K, 32, 1, 16K, 32, 1]$	1.34e-08	$[8K, 32, 1, 4K, 16, 2]$	1.35e-08	0.40%
g721dec	$[16K, 8, 1, 2K, 8, 2]$	1.63e-08	$[8K, 8, 8, 4K, 16, 2]$	2.83e-08	73.24%
gsmdec	$[8K, 32, 1, 16K, 32, 1]$	1.24e-08	$[8K, 32, 1, 16K, 32, 1]$	1.24e-08	0.00%
jpegdec	$[8K, 32, 1, 32K, 8, 1]$	1.72e-08	$[32K, 8, 1, 32K, 8, 1]$	1.91e-08	11.00%
jpegenc	$[16K, 32, 1, 32K, 32, 1]$	1.50e-08	$[16K, 32, 1, 16K, 32, 2]$	1.55e-08	3.08%
mpeg2dec	$[32K, 8, 1, 2K, 8, 1]$	1.38e-08	$[16K, 32, 2, 2K, 8, 1]$	1.48e-08	7.79%

Table 9.4. **Comparison between the full search optimal configurations (a_{opt}) and the configurations obtained with the sensitivity optimizer ($\hat{a}_{opt}$).**

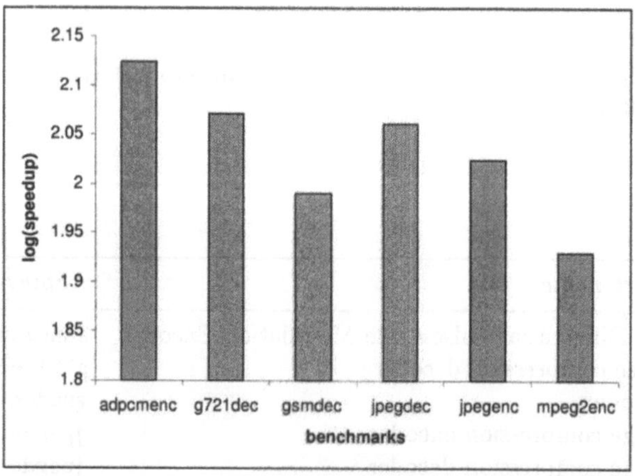

Figure 9.14. **The optimization time speedup obtained by the sensitivity optimizer is, in the average, of two orders of magnitude.**

4. The μSPARC architecture: a case study

The second case study is based on the SunMicrosystems μ SPARC2-based architecture to be optimized by the application of the sensitivity-based local search algorithm. The target system is composed of the following components:

- A 100MHz MicroSPARC2 Processor Core (operating at 3.3V and without on-chip caches).
- Separated and configurable I- and D-caches with one-cycle hit time. The speed of the processor-to-memory bus is 100MHz. The power model for the caches has been derived from [125].
- A 32Mbyte DRAM composed of 16 × 16-Mbit blocks and characterized by a 7-cycle latency. The power model for this memory has been derived from [213]. The speed of the bus cache-DRAM is 100MHz.

The cache energy model corresponds to $0.8\mu m$ $CMOS$ technology, however the equations in the analytical cache energy model can be easily modified to reflect a more up to date process technology by simply changing the values of the capacitance parameters to account for the new technological and layout features.

Each instance of the virtual architecture has been described as a *6-tuple* $[c_i, b_i, v_i, c_d, b_d, v_d] \in \mathcal{A} = S_{c_i} \times S_{b_i} \times S_{v_i} \times S_{c_d} \times S_{b_d} \times S_{v_d}$ where:

- S_{c_i}, S_{c_d} are the ordered sets of the possible sizes of the I- and D-caches (e.g. {2KB, 4KB, 8KB,16KB, 32KB, 64KB}).
- S_{b_i}, S_{b_d} are the ordered sets of the possible block sizes of the I- and D-caches (e.g. {4B, 8B, 16B, 32B})
- S_{v_i}, S_{v_d} are the ordered sets of the possible associativity values of the I- and D-caches (e.g. {1, 2, 4, 8})

The architecture has been explored by using a SPARC executable system module, called **MEX**, that consists of the SPARC V8 Instruction Set Simulator with a configurable memory architecture. **MEX** exploits the Shade [53] library to trace the memory accesses generated by a program executed by the SPARC V8 and simulates the target memory architecture to obtain accurate memory access statistics such as:

- Number of accesses to the memory hierarchy (on both I- and D-buses);
- Average cache miss rate (on both I- and D-buses);

- Address sequentiality (on both I and D address buses) [160];
- Bus transition activities (on both I and D address buses);
- Delay D (average clock cycles per instruction, measured in terms of $[CPI]$);
- Energy E dissipated by the architecture during program execution measured in terms of *Joule per Instruction* $[JPI]$.

The benchmarks used to validate the model are part of the Mediabench suite [150] (see Appendix 2). The benchmarks have been executed (with the standard options provided in their makefile) by **MEX** to provide system statistics.

To provide the energy-delay trend for the target system, Figure 9.15 reports the scatter plot of the energy and the delay for each point in the architectural space (composed of 9126 points) for the `gsm decoder` benchmark. In this picture, we can observe the system with the absolute minimum delay (point A), with the minimum energy (point B) and with the minimum EDP value (point C). The EDP optimal point is, with respect to the most performant configuration A, 1.5% slower but saves as much as 7.9% in energy consumption while, with respect to the less energy consuming configuration B, C is 1.9% faster and 1.3% more energy consuming. Note that, going from A to C (B to C), the percentage of energy (performance) savings are always greater than performance (energy) losses.

Figure 9.16 reports the scatter plot of the Energy and the Delay for the `mesa` benchmark. In this case, there is a substantial tradeoff to be exploited since the EDP optimal point is, with respect to the most performant configuration A, 4.8% slower but saves 31.8% in energy consumption. Finally, with respect to the less power consuming configuration B, the optimal EDP point is 9.4% faster and only 2.6% more power consuming.

Let us consider more specifically the benchmark `mesa` in order to better understand the behavior of the system in terms of delay and energy metrics. Figures 9.17 and 9.18 show the delay and the energy-delay product in the neighborhood of the optimal configuration of the system for the benchmark `mesa` ($a_{opt} = [16384, 16, 4, 16384, 4, 4]$) by varying the cache sizes and block sizes.

Note that, the increase of the cache size while keeping a constant block size tends to minimize the delay since more blocks are present in the cache and the miss probability decreases. However, increasing the cache size increases also the energy consumption and this implies a trade-off between energy consumption and performance. This is clearly

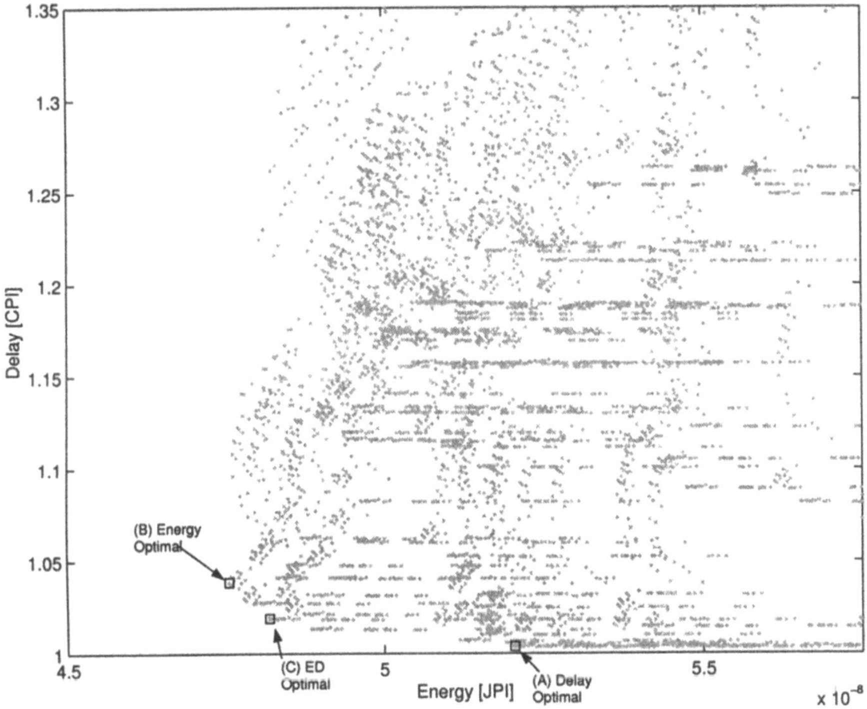

Figure 9.15. Normalized delay $[CPI]$ vs. energy $[JPI]$ scatter plot for the gsm decoder benchmark.

shown in the upper part of Figure 9.18 where the energy-delay curves are convex.

Figures 9.19 and 9.20 show the delay and the energy-delay product in the neighborhood of the optimal configuration of the system for the benchmark `mesa` by varying the cache sizes and set associativity of both caches. By increasing the cache size with a constant associativity increases the energy consumption and reduces delay, leading, also in this case, to a trade-off on the EDP.

Figures 9.21 and 9.22 show the delay and the energy-delay product in the neighborhood of the optimal configuration of the system for the benchmark `mesa` by varying the block sizes and associativity of both caches. Note that an increase in the block size with a constant associativity increases the delay (since data transfer time from the main memory becomes longer). However, larger associativity implies less delay, since it reduces cache misses. Furthermore, for the I-cache, increasing the cache block size leads also to an energy reduction, with fixed associativity and

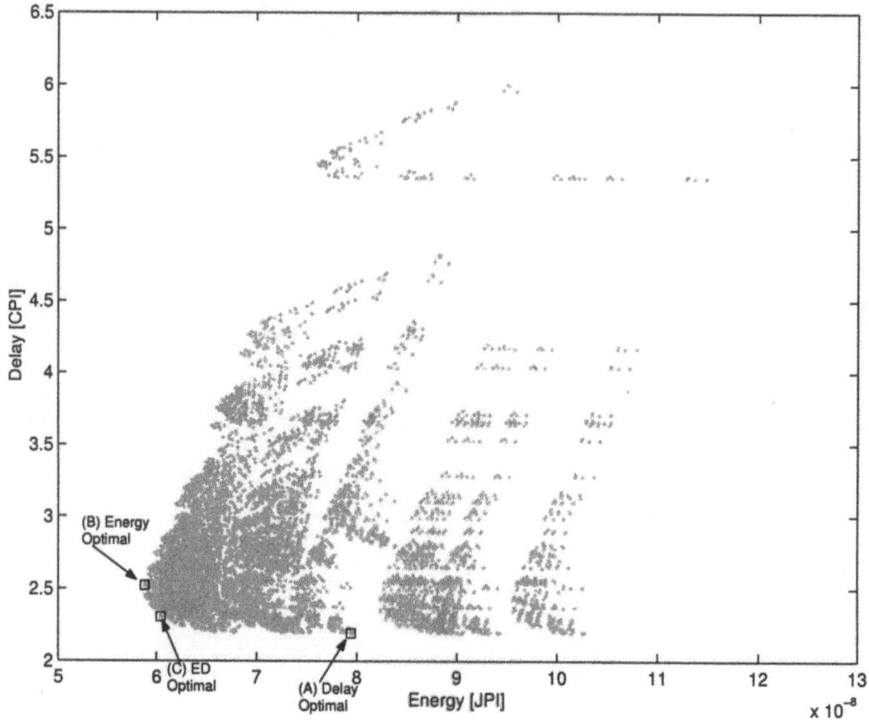

Figure 9.16. Normalized delay $[CPI]$ vs. energy $[JPI]$ scatter plot for the mesa benchmark.

fixed I-cache size, the number of cache misses is reduced. In this case, since the power model of the cache is very sensitive to the number of cache blocks, it is possible to find an optimal block size that minimizes both energy consumption and delay. Concerning the D-cache, the very high rate at which the delay increases with respect to the block size makes negligible the advantages obtained through the reduction of the cache complexity.

4.1 Application of the Tuning Phase

The tuning phase is based on the execution of the MediaBench suite [150] (see Appendix 2). Among the Mediabench, we selected a sub-set of control-intensive and data-intensive applications (shown in figure 9.5) to be used as reference application set $\mathcal{H}$ to characterize the sensitivity of the system.

Table 9.6 shows the optimal parameter values identified for the reference benchmarks and their sensitivity indexes. The last row (AVG)

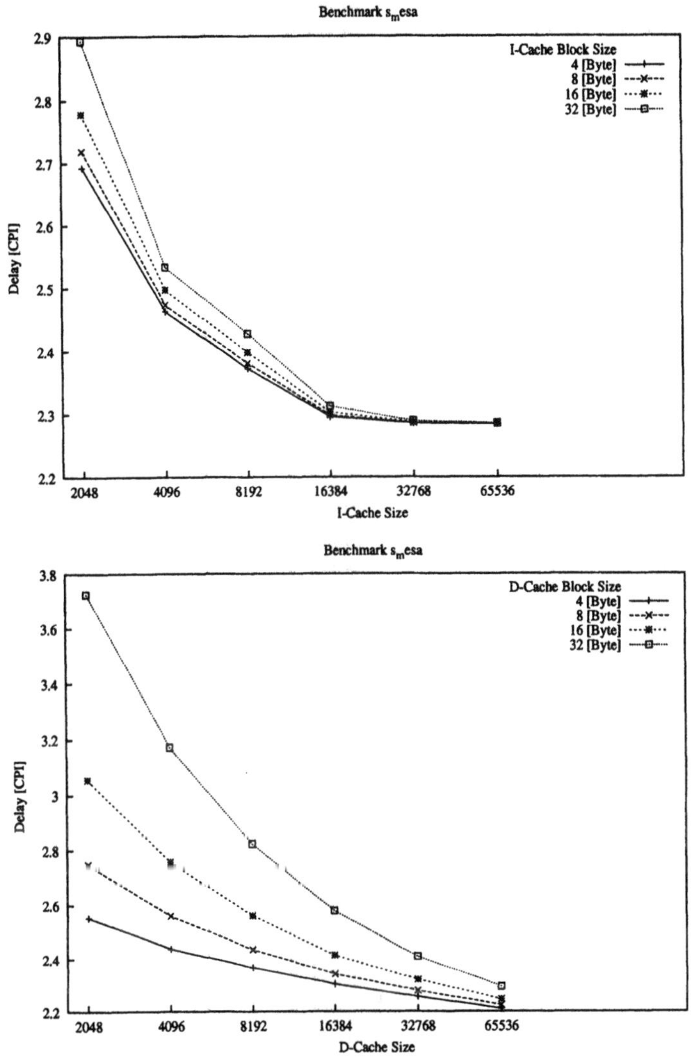

Figure 9.17. Delay for *mesa* benchmark, by varying cache sizes and block sizes

indicates the average sensitivity value associated with the parameter as well as the optimum feasible value of the parameter, i.e., the feasible value of the parameter closest to the average of the optimal values.

If we consider the average sensitivity, as Figure 9.23 shows, the most important parameters are in order, the I-cache size (c_i), the I-cache associativity (v_i), the D-cache size (c_d), the D-cache associativity (c_d) and the I- and D-cache blocks (b_i and b_d). This order will be used by the sensitivity optimizer to efficiently optimize any new application for the given architecture.

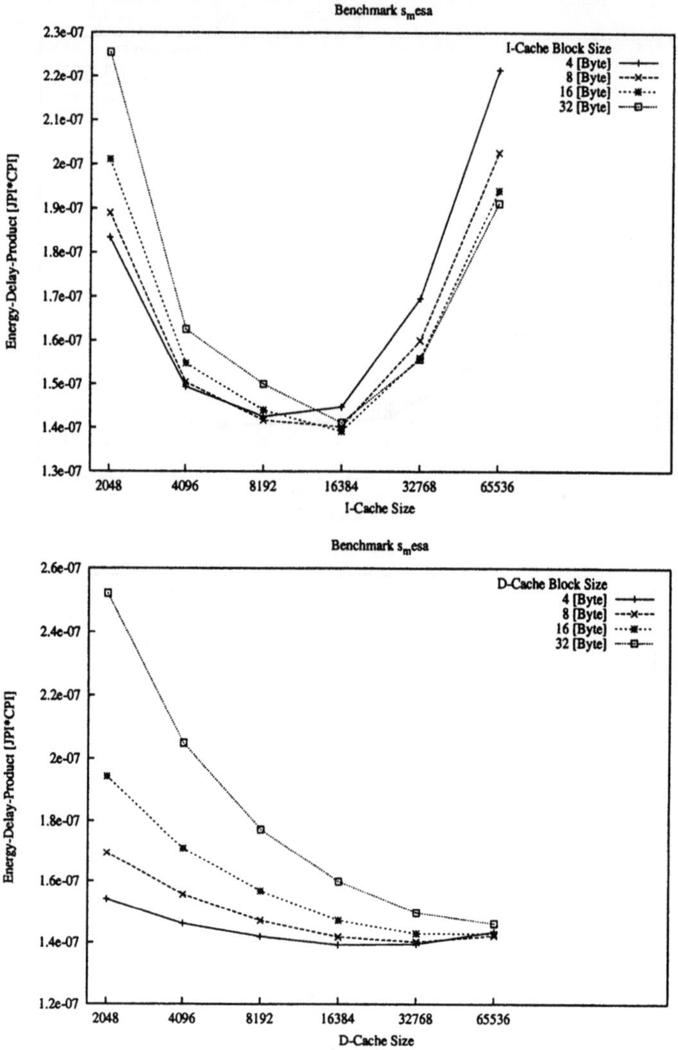

Figure 9.18. ED product for *mesa* benchmark, by varying cache sizes and block sizes

4.2 Sensitivity-Based Optimization

To validate the methodology, we applied the sensitivity based optimization to a completely different sub-set of Mediabench, with respect to the subset considered for the tuning phase. The sensitivity values used for the optimization are those identified in the previous section. Table 9.7 reports the applications chosen for the analysis.

Figure 9.24 shows the behavior of the sensitivity-based optimizer in the energy-delay space. As shown, the algorithm starts from a point characterized by high energy and delay values and tends to optimize

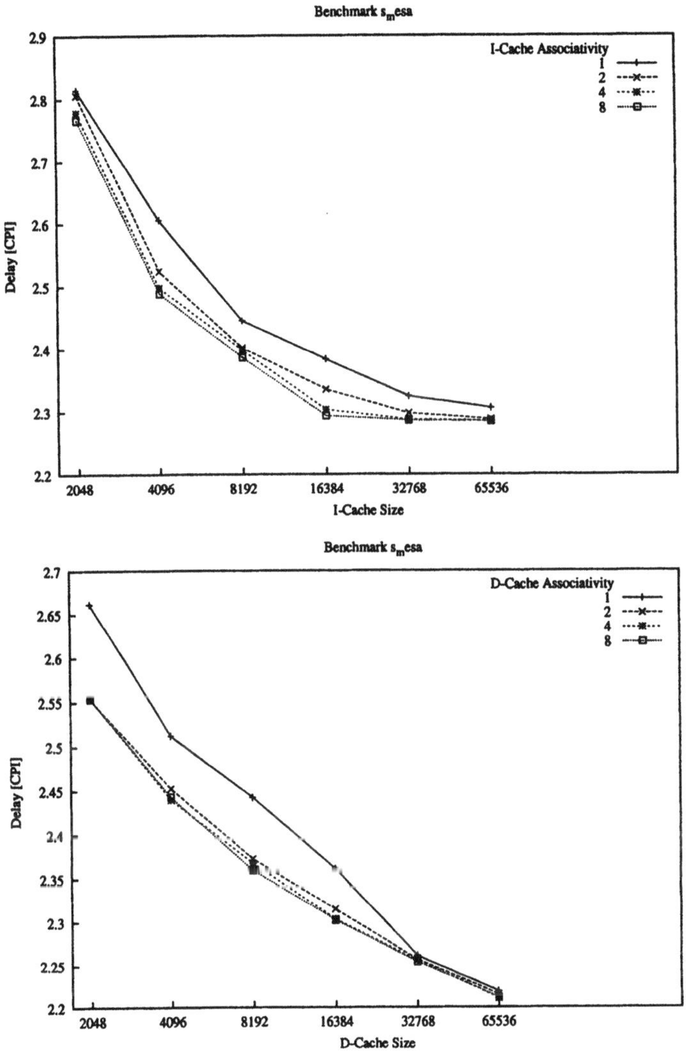

Figure 9.19. Delay for *mesa* benchmark, by varying cache sizes and associativity

energy and delay simultaneously. In this case, it reaches first a best delay configuration and, from this point, it begins to consider all the points of the pareto curve until it determines the best energy-delay configuration.

Table 9.8 shows the best configurations estimated by the sensitivity optimizer with respect to those found with a full search over the entire design space. The table shows also the percentage errors on the energy-delay product between the real and the estimated optimum. Note that for all benchmarks but three, the configuration found by the heuristic

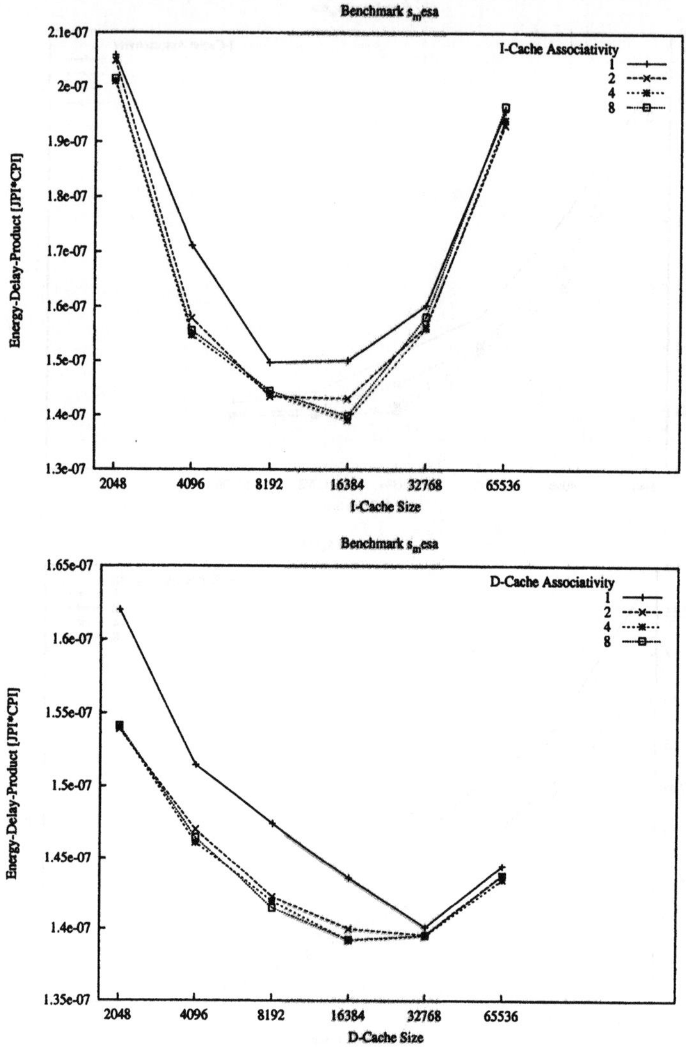

Figure 9.20. ED product for *mesa* benchmark, by varying cache sizes and associativity

coincides with the full search one. For the remaining benchmarks, the error on the Eenergy-Delay is always less than 2%.

Figure 9.25 reports the speedup in terms of number of simulations needed by the sensitivity optimizer. The average ratio between the number of simulations performed by the full search and the number of simulations performed by the sensitivity optimizer is always two orders of magnitude larger.

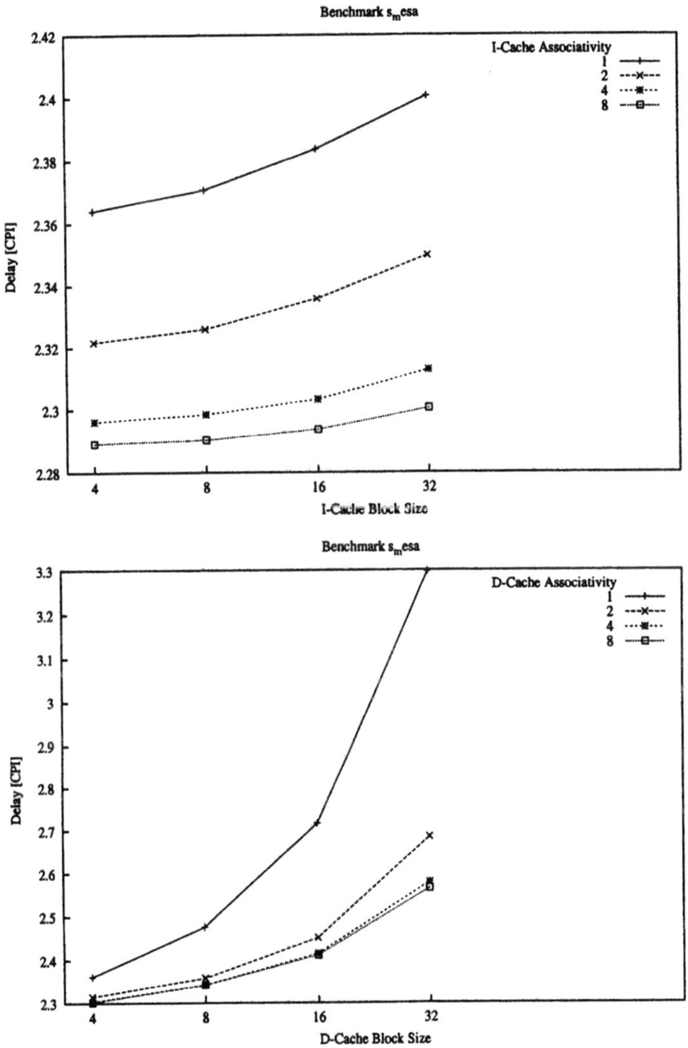

Figure 9.21. Delay for *mesa* benchmark, by varying block sizes and associativity

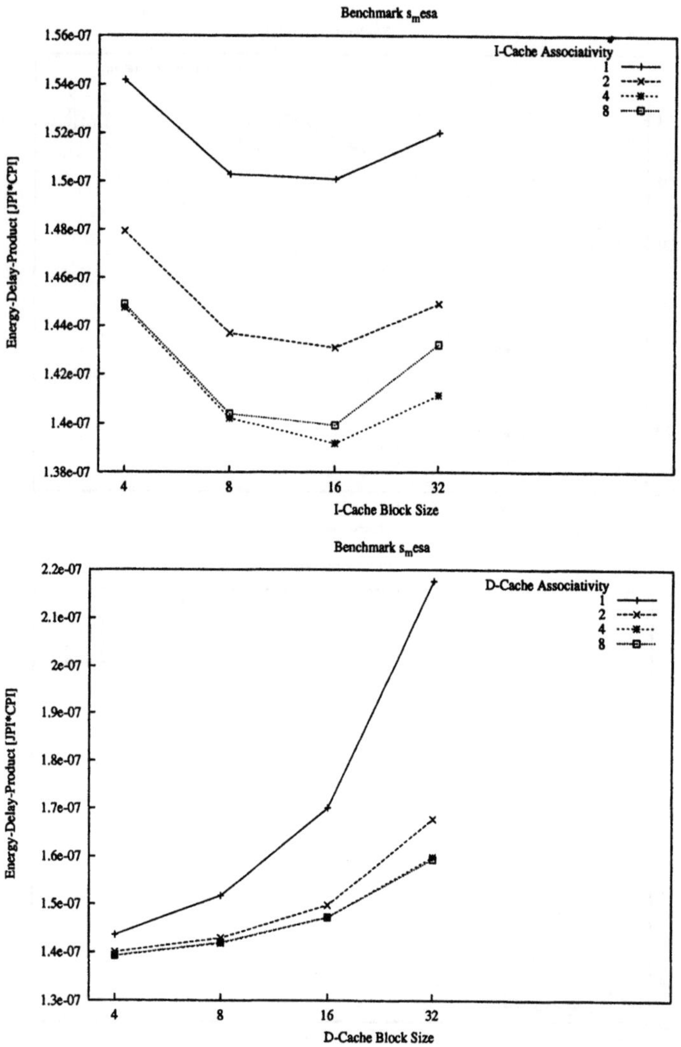

Figure 9.22. ED product for *mesa* benchmark, by varying block sizes and associativity

Benchmark name	Abbreviation
G.721 Voice Compression enc. and dec.	`g721enc,g72dec`
Ghostscript	`gsdec`
Mesa 3D Library	`mesa`
MPEG decoder	`mpegdec`
PGP digital signature and encryption	`pgp`
RASTA speech recognition	`rasta`

Table 9.5. The set of benchmarks used for the sensitivity analysis.

Bench.	c_i	$\sigma(c_i)$	b_i	$\sigma(b_i)$	v_i	$\sigma(v_i)$	c_d	$\sigma(c_d)$	b_d	$\sigma(b_d)$	v_d	$\sigma(v_d)$
g721dec	8*K*	106.7%	16	0.8%	2	41.3%	8*K*	1.2%	8	0.3%	2	0.5%
g721enc	8*K*	118.9%	16	0.8%	2	42.4%	8*K*	1.2%	8	0.2%	2	0.5%
gsdec	8*K*	6.2%	8	2.5%	8	0.3%	8*K*	2.7%	4	1.8%	2	8.8%
mesa	16*K*	12.1%	16	1.4%	4	2.8%	16*K*	2.0%	4	1.0%	4	0.6%
mpegdec	8*K*	8.6%	8	1.8%	4	2.3%	8*K*	2.4%	4	0.9%	4	0.6%
pgp	8*K*	45.7%	16	0.6%	2	5.9%	16*K*	15.4%	4	0.6%	2	1.4%
rasta	32*K*	20.4%	16	3.1%	8	3.0%	64*K*	17.7%	16	1.0%	4	4.9%
AVG	16*K*	51.1%	16	1.6%	4	15.9%	16*K*	6.8%	8	0.8%	2	2.8%

Table 9.6. The set of optimal parameters and sensitivity indexes for the chosen set of reference applications.

Benchmark name	Abbreviation
Adaptive Differential PCM Enc. and Dec.	`adpcmenc,adpcmdec`
EPIC image encoder	`epic`
GSM encoder and decoder	`gsmenc,gsmdec`
Jpeg image compression encoder and decoder	`jpegenc,jpegdec`
PEGWIT public key encryption	`pegwit`

Table 9.7. The set of benchmarks used for the validation of the sensitivity approach

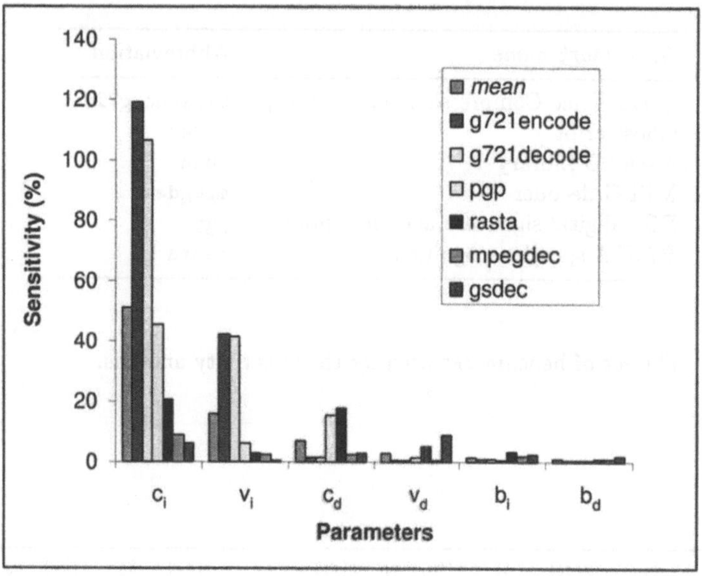

Figure 9.23. The distribution of the sensitivity for the set of the architectural parameters ordered by their mean sensitivity

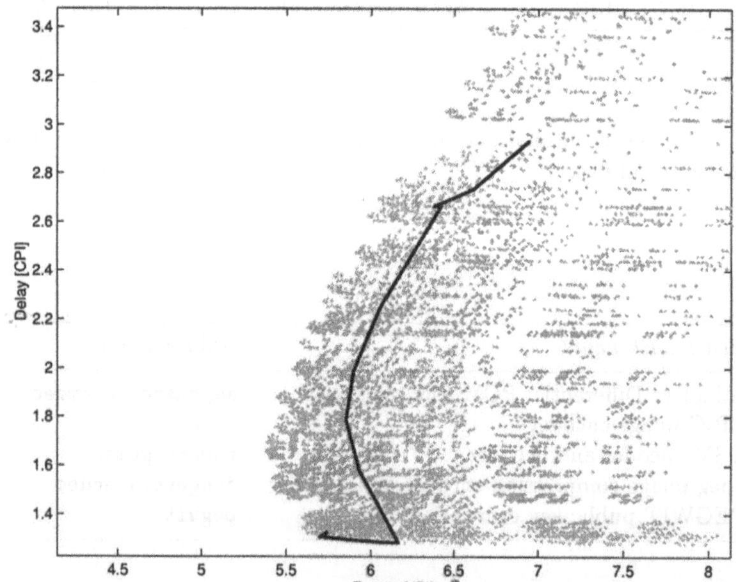

Figure 9.24. The path followed by the sensitivity-based algorithm when optimizing the **jpeg** encoder application on the **microSPARC** system

Bench.	a_{opt}	$ED(a_{opt})$	$\hat{a}_{opt}$	$ED(\hat{a}_{opt})$	*error*
adpcmdec	$[4K, 4, 4, 8K, 4, 2]$	5.47e-08	$[4K, 4, 4, 8K, 4, 2]$	5.47e-08	0.00%
adpcmenc	$[4K, 4, 4, 8K, 4, 2]$	5.41e-08	$[4K, 4, 4, 8K, 4, 2]$	5.41e-08	0.00%
epic	$[2K, 8, 2, 32K, 4, 4]$	1.03e-07	$[2K, 8, 2, 16K, 4, 2]$	1.03e-07	0.07%
gsmdec	$[8K, 8, 8, 16K, 32, 2]$	4.91e-08	$[8K, 8, 8, 32K, 32, 1]$	4.99e-08	1.67%
gsmenc	$[16K, 16, 2, 8K, 16, 1]$	5.45e-08	$[16K, 16, 2, 8K, 16, 1]$	5.45e-08	0.00%
jpegdec	$[4K, 4, 4, 32K, 16, 4]$	6.46e-08	$[4K, 4, 4, 32K, 16, 4]$	6.46e-08	0.00%
jpegenc	$[4K, 8, 4, 64K, 8, 1]$	7.42e-08	$[4K, 8, 4, 64K, 8, 1]$	7.42e-08	0.00%
mpegenc	$[2K, 4, 4, 4K, 4, 4]$	5.19e-08	$[4K, 16, 2, 8K, 8, 4]$	5.24e-08	0.91%
pegwit	$[8K, 16, 2, 64K, 4, 1]$	9.05e-08	$[8K, 16, 2, 64K, 4, 1]$	9.05e-08	0.00%
unepic	$[2K, 8, 2, 64K, 8, 4]$	2.71e-07	$[2K, 8, 2, 64K, 4, 2]$	2.72e-07	0.22%

Table 9.8. Comparison between the full search optimal configurations (a_{opt}) and the configurations derived with the sensitivity optimizer ($\hat{a}_{opt}$).

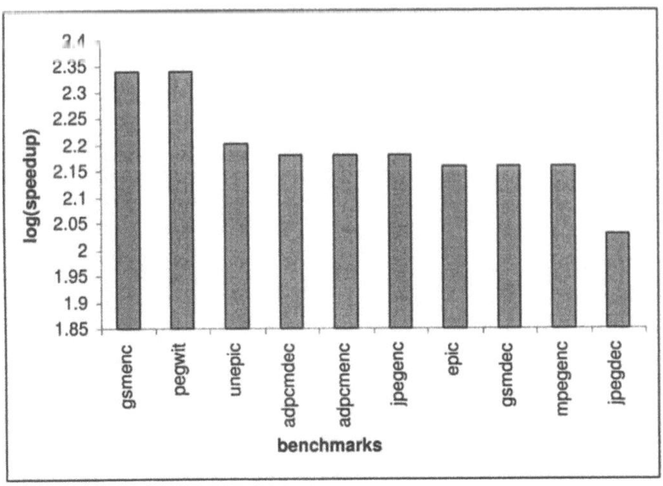

Figure 9.25. The optimization time speedup obtained by the sensitivity optimizer is always greater than two orders of magnitude.

5. Conclusions

This chapter addressed the problem of design space exploration of embedded system architectures considering performance and power consumption as the most relevant constraints. In particular, an exploration methodology to reduce simulation time while preserving acceptable design accuracy has been proposed and experimentally assessed by considering the design of the memory subsystem of two real-world embedded systems. Experimental results have shown a speed-up in simulation time of almost two orders of magnitude and an average distance from the optimal configuration below 16%. Furthermore, this methodology allows the designer to save analysis time since the number of configurations to be compared is significantly reduced.

Beyond the evaluation of the impact of the variation of cache-related parameters from the power-performance joint perspective, the next step of our work will aim at estimating the effects of power-aware compiler optimizations on the Energy*Delay metric at the system-level also for superscalar architectures.

Chapter 10

CONCLUSIONS AND FUTURE WORK

The semiconductor market is characterized by an increasing demand for low power system-on-a-chip solutions imposing new challenges mainly related to the management of the complexity, and the support for the specification, design, analysis and validation phases.

Given the complexity of envisioned systems, the ideal approach could be to involve the designer in architectural space exploration and high-level optimization, while synthesis (hardware and software) would be best carried out by CAD tools. Such a solution would exploit the best characteristics of human designers - namely, creativity and flexibility - and of automatic tools, i.e., efficiency and accuracy.

Given the ever-increasing importance of power as a design constraint, satisfactory architecture-space exploration can be carried out only if fast and accurate tools for power estimation and optimization are available already at the higher abstraction levels.

The main contribution of this thesis is the introduction of innovative power estimation and optimization methodologies to support the design of low power embedded systems based on high-performance VLIW microprocessors.

The proposed power estimation techinques address the instruction-level as well as the system-level power consumption of VLIW-based architectures. Such techniques allow a fast and accurate power estimation framework by reducing the characterization complexity proper of all instruction level models that take into account inter-instruction effects. The proposed techniques have been used to estimate the power consumption of an industrial VLIW architecture by providing a significant speed-up of the estimation process and a reasonable accuracy.

V. Zaccaria et al. (eds.),
Power Estimation and Optimization Methodologies for VLIW - based Embedded Systems, 181–184.

The proposed power optimization techniques address the microarchitectural-level as well as the system-level. Two main optimization techniques have been studied: the definition of register file write inhibition schemes that exploit the forwarding paths, and the definition of a design exploration framework for an efficient fine-tuning of the configurable modules of an embedded system. Both techniques have been validated on industrial case studies showing significant power savings.

In the following sections, we outline the ongoing and future research directions starting from the techniques and methodologies proposed in this thesis.

1. Ongoing and future research on power estimation

Fast and accurate power optimization and estimation tools are needed for the design phases of complex digital systems. Such efficiency can be achieved by raising the level of abstraction to the system-level, where the granularity of the optimizations is much coarser.

In the near future, our main effort will be directed to investigating the extension of the estimation engines presented in this thesis to extensively support system-level exploration. In this scenario, the following directions have been devised:

Multicluster support. The VLIW power model proposed in this thesis aims at estimating the energy consumption of only one VLIW core. However, multiple *clusters* can be configured in a *multi-cluster* architecture in which all the clusters share the same program counter and the same instruction cache. Besides, each cluster has its own, reserved, register file and data-cache and explicit data movements operations must be used to transfer data from one register file to another. Moreover, data coherency protocols must be used to maintain coherent data caches characterized by multiple and mutually independent access ports. We are currently investigating how the power model for a multi-cluster system can be built starting from the power model of the single clusters and how the complexity of such a model can be maintained under a reasonable level.

Compiler/operating system support. The power models proposed in this thesis give to the compiler and the OS a direction to influence the power consumption of a schedule of instructions by modifying the order of instruction execution. We are currently investigating some compiler level techniques that exploit the power models to modify the properties of the scheduled code such as minimization of the power consumption, flattening of the power consumption (under a fixed power level), minimization of the variation rates of power consumption (to extend battery life). Voltage scaling techniques can also be included in the power model to guide OS/compiler towards system-level power management.

Bus topologies and bus encoding techniques. We are currently working on the power estimation of complex VLIW and superscalar architectures connected by an arbitrary interconnection topology, in particular for multi-clusterd architectures. For each bus a different bus encoding technique can be devised in order to minimize the power consumption. The power estimation techniques are useful to guide the choice of low power bus encodings expecially for off-chip buses that are characterized by a high switching capacitance.

2. Ongoing and future research on power optimization

Power optimization can be applied at all abstraction levels starting from the high-level design specification down to gate-level techniques and beyond. However, significant power savings can be achieved by operating at the highest abstraction levels, where many degrees of freedom can be exploited by the designer. Our work is directed towards the investigation of high-level innovative power optimization techniques for microprocessor based systems. In this view, the following directions for the research have been devised:

System-level power exploration. The target is to define novel methodologies for the design space exploration of complex systems, with different inter-connection topologies within the modules and arbitrary large space of parameters (such as bus encoding techniques or microprocessor internal buffers sizing). Among the various parameters to be taken into account there are:

- Multiple levels of cache hierarchies
- Multiple VLIW clusters
- Bus protocols
- Compiler optimization parameters

These techniques are based on heuristic exploration algorithms such as *random search* algorithms, since a full search of the design space is unfeasible.

Compiler/application level power optimization. As noted before, a viable direction for the research is to modify the properties of the scheduled code for the minimization of the power consumption, the equalization of the power consumption (below a fixed power level) and the minimization of the variation rates of power consumption (to extend battery life). Voltage scaling techniques can also be envisioned to implement OS/compiler toward system-level power management.

Low power finite state machines. Significant power savings can be achieved by operating on the control modules of the processor. This is usually described as a set of interacting finite state machines that dissipate power during each state transition. We are currently taking into consideration the modification of the state encoding for such FSMs, to minimize power consumption. Moreover we are also investigating the application of clock-gating techniques to the sequential elements of the processor.

Low power register forwarding. The basic idea discussed in chapter 8 can be extended by thinking at the set of pipeline micro-registers as a new level in the memory hierarchy, i.e., the microregisters-level. This new memory-level can avoid wasting space in the *RF* for registers whose liveness length is very short and therefore reducing register spilling.

The compiler, whenever short-lived variables are found such that use of forwarding is applicable and that the conditions specified below are satisfied, does not reserve registers in the *RF* for such variables. The *RF* space thus is effectively "increased" as far as compiler use is concerned; register spilling (and ensuing cache traffic) is reduced.

In conclusion, ongoing research is directed towards the solution of some of the most crucial problems affecting the current development of increasingly complex embedded systems, by introducing innovative power estimation and optimization for embedded SoC.

The Mediabench suite

The Mediabench suite [150] is a set of multimedia benchmarks that can be considered a *de-facto* standard for multimedia applications. All the applications are public domain and derive from image processing, communication and DSP applications. The suite is composed of the following applications:

G.721 Voice Compression encoder and decoder: These applications consist of the reference implementations of the CCITT (International Telegraph and Telephone Consultative Committee) G.711, G.721, and G.723 voice compressions algorithms.

Ghostscript: This application is an interpreter for the PostScript language characterized by a very high I/O activity but no graphical display.

Mesa: Mesa is a 3D graphics library very similar to OpenGL. Among the available applications using this library we selected *osdemo*, which executes a standard rendering pipeline.

MPEG decoder: This is an implementation of an MPEG2 decoder, a standard for high-quality digital video compression, based on the inverse discrete cosine transform.

PGP encoder: PGP is digital signature and encryption application based on "message digests", i.e, 128-bit cryptographically strong one-way hash value of the message. Data encryption is based on the RSA asymmetric encryption scheme.

RASTA: This application performs speech recognition by means of speech analysis techniques such as PLP, RASTA, and Jah-RASTA.

ADPCM encoder and decoder: Adaptive differential pulse code modulation filters for audio coding.

EPIC encoder and decoder: Image compression/decompression fixed point tools based on a bi-orthogonal wavelet decomposition, combined with a run-length/Huffman entropy coder.

GSM encoder and decoder: European GSM 06.10 standard speech transcoders based on residual pulse excitation/long term prediction coding at 13 kbit/s.

JPEG encoder and decoder: Standardized lossy compression method for full-color and gray-scale images.

MPEG encoder: This is an implementation of an MPEG2 encoder, a standard for high-quality digital video compression, based on the inverse discrete cosine transform.

PEGWIT: Public key encryption and authentication application based on the elliptic curves theory.

References

[1] K. Keutzer, S. Malik, A. R. Newton, J. Rabaey, and A. Sangiovanni-Vincentelli, "System level design: Orthogonolization of concerns and platform-based design," *IEEE Transactions on Computer-Aided Design of Integrated Circuits and Systems*, vol. 19, no. 12, pp. 1523–1543, December 2000.

[2] G. De Micheli and M. Sami, *Hardware-software codesign*, Kluwer Academic Publishers, 1996.

[3] D. Gajski, N. Dutt, A. Wu, and S. Lin, *High-Level Synthesis, Introduction to Chip and System Design*, Kluwer Academic Publishers, 1994.

[4] J. Rozenblit and K. Buchenrieder, *Codesign, Computer-Aided Software/Hardware Engineering*, IEEE Press, 1995.

[5] J. van den Hurk and J. Jess, *System-Level Hardware-Software Codesign*, Kluwer Academic Publishers, 1998.

[6] H. De Man, I. Bolsens, B. Lin, K. Van Rompaey, S. Vercauteren, and D. Verkest, "Co-design of dsp systems," In G. De Micheli and M. Sami, editors, Hardware-Software Co-Design, Kluwer Academic Publishers, Norwell, MA, 1996.

[7] F. Balarin, P. Giusto, A. Jurecska, C. Passerone, E. Sentovich, B. Tabbara, M. Chiodo, H. Hsieh, L. Lavagno, A. Sangiovanni-Vincentelli, and K. Suzuk, "Hardware-software co-design of embedded systems: The polis approach," Kluwer Academic Publishers, 1997.

[8] J. Staunstrup and W. Wolf, *Hardware/Software CoDesign: Principles and Practice*, Kluwer Academic Publishers, 1997.

[9] G. De Micheli, "Hardware–software codesign: Extending cad tools and techniques," *IEEE Computer*, pp. 85–87, Jan. 1993.

[10] Virtual Socket Interface Alliance, *System Level Design Model Taxonomy*, VSI Alliance Inc., 1998.

[11] L. Nagel, "SPICE2: A Computer Program to Simulate Semiconductor Circuits," Tech. Rep. ERL-M520, UCBerkeley, May 1975.

[12] N. Weste and K. Eshraghian, *Principles of CMOS VLSI Design*, Addison Wesley, 1988.

[13] G. Nifong, D. Erdman, and D. Rose, "CAzM: Circuit analyzer with macromodeling," Tech. Rep. ERL-M520, MCNC, june 1990.

[14] A. Deng, "Power analysis for cmos/bicmos circuits," in *Proceedings of International Symposium on Low-Power Electronics and Design (ISLPED)*, 1994, pp. 3–8.

[15] A. Salz and Mark Horowitz, "IRSIM: An incremental MOS switch-level simulator," in *Proceedings of Design Automation Conference 89*, 1989, pp. 173–178.

[16] DAS Subcommittee, "Ieee standard vhdl language reference manual," 1987.

[17] IEEE Computer Society, *IEEE Standard Hardware Description Language Based on the Verilog Hardware Description Language*, IEEE Computer Society Press, Piscataway, USA, 1996.

[18] Open SystemC Initiative, "SYSTEM C Version 1.0," User Guide, 2000.

[19] Synopsys, Inc., "VSS Reference. (Version 3.3a)," Reference Manual, April 1995.

[20] Chronologic, "VCS Reference Manual, version 2.0," Reference Manual, 1993.

[21] Cadence Design Systems Inc., "Verilog-xl reference manual (v. 2.1)," December 1994.

[22] H. Tomiyama, A. Halambi, P. Grun, N. Dutt, and A. Nicolau, "Architecture description languages for system–on–chip design," in *In Proc. APCHDL*, Fukuoka, Japan, October 1999.

[23] S. Bashford, U. Bieker, B. Harking, R. Leupers, P. Marwedel, A. Neumann, and D. Voggenauer, "The mimola language version 4.1," Tech. Rep., Lehrstul Informatik XII, University of Dortmund, Sept. 1994.

[24] UDL/I Committee, "Udl/i language reference manual version 2.1.0a," 1994.

[25] ASTEM RI, "Udl/i logic synthesis system user's guide ver. 1.1.0," 1995.

[26] J. Gyllenhaal, "A machine description language for compilation," M.S. thesis, Department of Electrical and Computer Engineering, University of Illinois, Urbana IL, Sept. 1994.

[27] B. Goldberg, H. Kim, V. Kathail, and J. Gyllenhaal, "The trimaran compiler infrastructure for instruction level parallelism research," Tech. Rep., Hewlett-Packard Laboratories, University of Illinois, NYU, 1998.

[28] John C. Gyllenhaal, Wen mei W. Hwu, and B. Ramakrishna Rau, "Optimization of machine descriptions for efficient use," in *International Symposium on Microarchitecture*, 1996, pp. 349–358.

[29] V. Kathail, M. Schlansker, and B. Rau, "Hpl playdoh architecture specification: Version 1.0," Tech. Rep. HPL-93-80, Hewlett-Packard Laboratories, Palo Alto, CA 94303, February 1994.

[30] P. Paulin, C. Liem, T. May, and S. Sutarwala, "Flexware: A flexible software development environment for embedded systems," In P. Marwedel and G. Goossens, editors, Code Generation for Embedded Processors, Kluwer Academic Publishers, 1995.

[31] A. Appel, J. Davidson, and N. Ramsey, "The zephyr compiler infrastructure," Tech. Rep., University of Virginia, 1998.

[32] Norman Ramsey and Mary F. Fernández, "Specifying representations of machine instructions," *ACM Transactions on Programming Languages and Systems*, vol. 19, no. 3, pp. 492–524, May 1997.

[33] Norman Ramsey and Jack W. Davidson, "Machine descriptions to build tools for embedded systems," Tech. Rep. CS-98-02, Department of Computer Science, University of Virginia, 19, 1998.

[34] Mark W. Bailey and Jack W. Davidson, "A formal model of procedure calling conventions," in *Conference Record of POPL '95: 22nd ACM SIGPLAN-SIGACT Symposium on Principles of Programming Languages*, San Francisco, California, 1995, pp. 298–310.

[35] A. Halambi, P. Grun, V. Ganesh, A. Khare, N. Dutt, and A. Nicolau, "Expression: A language for architecture exploration through compiler /simulator retargetability," in *Proceedings of Design Automation and Test in Europe*, Munich, Germany, 1999.

[36] V. Zivojnovic, "Lisa - machine description language and generic machine model for hw/sw co-design," in *In Proc. of IEEE Workshop on VLSI Signal Processing*, 1996.

[37] C. Siska, "A processor description language supporting retargetable multipipeline DSP program development tools," in *In Proc. ISSS*, December 1998.

[38] T. Morimoto, K. Saito, H. Nakamura, T. Boku, and K. Nakazawa, "Advanced processor design using hardware description language aidl," in *In Proc. of ASPDAC*, 1997.

[39] Doug Burger, Todd M. Austin, and Steve Bennett, "Evaluating future microprocessors: The simplescalar tool set," Tech. Rep. CS-TR-1996-1308, University of Wisconsin, 1996.

[40] J. Veenstra and R. Fowler, "Mint: A front end for efficient simulation of shared-memory multiprocessors," January 1994.

[41] V. Pai, P. Ranganathan, and S. Adve, "Rsim: An execution-driven simulator for ilp-based shared-memory multiprocessors and uniprocessors," in *In Proceedings of the Third Workshop on Computer Architecture Education*, February 1997.

[42] M. Freericks, "The nmlmachine description formalism," Tech. Rep. TR-SM-IMP/DIST/08, TU Berlin CS Dept., 1993.

[43] A. Fauth, M. Freericks, and A. Knoll, "Generation of hardware machine models from instruction set descriptions," in *Proc. IEEE Workshop VLSI Signal Proc.*, Veldhoven (Netherlands), Oct. 1993, pp. 242–250.

[44] F. Lohr, A. Fauth, and M. Freericks, "Sigh/sim: An environment for retargetable instruction set simulation," Tech. Rep. 1993/43, Fachbereich Informatik, TU Berlin, 1993.

[45] Mark R. Hartoog, James A. Rowson, Prakash D. Reddy, Soumya Desai, Douglas D. Dunlop, Edwin A. Harcourt, and Neeti Khullar, "Generation of software tools from processor descriptions for hardware/software codesign," in *Design Automation Conference*, 1997, pp. 303–306.

[46] George Hadjiyiannis, Silvina Hanono, and Srinivas Devadas, "ISDL: An instruction set description language for retargetability," in *Design Automation Conference*, 1997, pp. 299–302.

[47] S. Hanono and S. Devadas, "Instruction selection, resource allocation, and scheduling in the aviv retargetable code generator," in *35th Design Automation Conference*, June 1998, pp. 510–515.

[48] J. Bell, "Threaded code," *Communications of ACM*, vol. 16, pp. 370–372, 1973.

[49] P. Magnusson, "Performance debugging using simics," Tech. Rep., Swedish Institute of Computer Science, 1995.

[50] Rok Sosic, "Dynascope: A tool for program directing," in *Proceedings of the Conference on Programming Language Design and Implementation (PLDI)*, New York, NY, 1992, vol. 27, pp. 12–21, ACM Press.

[51] D. A. Patterson and J. L. Hennessy, *Computer Architecture: A Quantitative Approach*, Morgan Kaufmann, San Mateo, CA, 2nd edition, 1996.

[52] R. Bedichek, "Talisman: Fast and accurate multicomputer simulation," in *in Proceedings of Sigmetrics 95/Performance 95*, May 1995, pp. 14–24.

[53] Bob Cmelik and David Keppel, "Shade: A fast instruction-set simulator for execution profiling," *ACM SIGMETRICS Performance Evaluation Review*, vol. 22, no. 1, pp. 128–137, May 1994.

[54] Mendel Rosenblum, Stephen A. Herrod, Emmett Witchel, and Anoop Gupta, "Complete computer system simulation: The SimOS approach," *IEEE parallel and distributed technology: systems and applications*, vol. 3, no. 4, pp. 34–43, Winter 1995.

[55] Alan Eustace and Amitabh Srivastava, "ATOM: A flexible interface for building high performance program analysis tools," in *Proceedings of the Winter 1995 USENIX Conference*, January 1995, pp. 303–313.

[56] Amitabh Srivastava and Alan Eustace, "ATOM: A system for building customized program analysis tools," *ACM SIGPLAN Notices*, vol. 29, no. 6, pp. 196–205, 1994.

[57] Amitabh Srivastava and David W. Wall, "A practical system for intermodule code optimization at link-time," *Journal of Programming Languages*, vol. 1, no. 1, pp. 1–18, December 1992.

[58] M. Smith, "Tracing with pixie," Tech. Rep., Center for Integrated Systems, Stanford University, Stanford, CA, April 1991.

[59] S. Mukherjee, "Wisconsin wind tunnel ii: A fast and portable parallel architecture simulator," in *In Workshop on Perf. Analysis and Its Impact on Design (PAID)*, June 1997.

[60] D.L. Barton, "System level design section of the industry standards roadmap," in *Proceedings of SLDL Workshop Introduction, First Int. Forum of Design Languages*, Lausanne, Switzerland, Sept. 1998, vol. 2, pp. 9–28.

[61] J. Buck, S. Ha, E. Lee, and D. Messerschmitt, "Ptolemy: A Framework For Simulating and Prototyping Heterogeneous Systems," *International Journal of computer Simulation, special issue on Simulation Software Development*, vol. 4, 1994.

[62] A. P. Chandrakasan and R. W. Brodersen, *Low-Power Digital CMOS Design*, Kluwer Academic Publishers, Boston-Dordrecht-London, 1995.

[63] K. Roy and S. C. Prasad, *Low-Power CMOS VLSI Circuit Design*, Wiley Interscience, 2000.

[64] H.J.M. Veendrick, "Short-circuit dissipation of static CMOS circuitry and its impact on the design of buffer circuits.," *IEEE Journal of Solid-State Circuits*, vol. SC-19, pp. 468–473, August 1984.

[65] N. Hedenstierna and K. Jeppson, "CMOS Circuits Speed and Buffer Optimization," *IEEE Transactions on Computer-Aided Design of Integrated Circuits and Systems*, vol. CAD-6, no. 2, pp. 270–280, March 1987.

[66] S. Vemuru and N. Scheinberg, "Short-Circuit Power Dissipation Estimation for CMOS Logic Gates," *IEEE Transactions on Circuits and Systems*, vol. CAS-41, no. 11, pp. 762–765, Nov. 1994.

[67] T. Nose, K.and Sakurai, "Analysis and future trend of short-circuit power," *IEEE Transactions on Computer-Aided Design of Integrated Circuits and Systems*,.vol. CAD-19, no. 9, pp. 1023–1030, Sept. 2000.

[68] L. Bisdounis, , and O. Koufopavou, "Short-circuit energy dissipation modeling for submicrometer CMOS gates," *IEEE Transactions on Circuits and Systems I: Fundamental Theory and Applications*, vol. CAS-47, no. 9, pp. 1350 - 1361, Sept. 2000.

[69] J. Rabaey and M. Pedram, *Low Power Design Methodologies*, Kluwer Academic Publishers, 1995.

[70] P. Landman, "Irsim-cap - modified version of irsim for better capacitance measurements," Tech. Rep., Univ. of Calif. Berkerley, 1995.

[71] R. Tjarnstrom, "Power dissipation estimate by switch level simulation," in *Proceedings IEEE International Symposium on Circuits and Systems*, may 1989, pp. 881–884.

[72] Kenneth P. Parker and Edward J. McCluskey, "Probabilistic Treatment of General Combinational Networks," *IEEE Transactions on Computers*, vol. 24, no. 6, pp. 668–670, 1975.

[73] L. Benini, G. De Micheli, E. Macii, M. Poncino, and S. Quer, "System-level power optimization of special purpose applications: The Beach solution," Aug. 1997.

[74] S. Akers and D. Diagrams, "Binary Decision Diagrams," *IEEE Transactions on Computers*, vol. C-27, pp. 509–516, 1978.

[75] T.-L. Chou, K. Roy, and S. Prasad, "Estimation of circuit activity considering signal correlations and simultaneous switching," in *International Conference on Computer-Aided Design*, Los Alamitos, Ca., USA, Nov. 1994, pp. 300–303, IEEE Computer Society Press.

[76] R. Marculescu, D. Marculescu, and M. Pedram, "Probabilistic Modeling of Dependencies During Switching Activity Analysis," *IEEE Transactions on Computer Aided Design of Integrated Circuits and Systems*, vol. 17, no. 2, pp. 73–83, Feb. 1998.

[77] R. Marculescu, D. Marculescu, and M. Pedram, "Composite sequence compaction for finite state machines using block entropy and higher-order markov models," in *Proceedings of ISLPED-97: ACM/IEEE International Symposium on Low-Power Electronics and Design*, Monterey, CA, 1997, pp. 190–195.

[78] A. Macii, E. Macii, M. Poncino, and R. Scarsi, "Stream synthesis for efficient power simulation based on spectral transforms," *IEEE Transactions on Very Large Scale Integration (VLSI) Systems*, vol. 9, no. 3, pp. 417–426, 2001.

[79] Enrico Macii, Massoud Pedram, and Fabio Somenzi, "High level power modeling, estimation, and optimization," *IEEE Transactions on CAD of integrated circuits and systems*, vol. 17, no. 11, pp. 1061–1079, Nov 1998.

[80] S. Powell and P. Chau, "Estimating power dissapation of vlsi signal processing chips: The pfa technique," in *Proc. of VLSI Signal Processing IV*, 1990, pp. 250–259.

[81] P. Landman and J. Rabey, "Architectural Power Analysis: The Dual Bit Type Method," *IEEE Trans. on VLSI Systems*, vol. 3, no. 2, pp. 173–187, June 1995.

[82] P. Landman and J. Rabaey, "Activity-Sensitive Architectural Power Analysis," *IEEE Transactions on CAD of Integrated Circuits and Systems*, vol. 15, no. 6, June 1996.

[83] Subodh Gupta and Farid N. Najm, "Power Macromodeling for High Level Power Estimation," *IEEE Transactions on VLSI Systems*, vol. 8, no. 1, pp. 18–29, 2000.

[84] Q. Wu, Q. Qiu, M. Pedram, and C. Ding, "Cycle-accurate Macromodels for RT-level power analysis," *IEEE Trans. on Very Large Scale Integration (VLSI) Systems*, vol. 6, pp. 520–528, Dec. 1998.

[85] Q. Wu, Q. Qiu, M. Pedram, and C. Ding, "Cycle-accurate macromodels for rt-level power analysis," in *Proc. International Symposium on Low Power Electronics and Design*, 1997, pp. 125–130.

[86] C. Hsieh, Q. Wu, C. Ding, and M. Pedram, "Statistical sampling and regression analysis for rt-level power evaluation," in *Proc. Int'l Conf. on CAD*, 1996 1996, pp. 583–588.

[87] Luca Benini, Giovanni de Micheli, Enrico Macii, Massimo Poncino, and Riccardo Scarsi, "A multilevel engine for fast power simulation of realistic input streams," *IEEE Transactions on CAD of Integrated Circuits and Systems*, vol. 19, no. 4, pp. 459–472, Apr. 2000.

[88] Michael Eiermann and Walter Stechele, "Novel modeling techniques for rtl power estimation," in *Proceedings of the 2002 international symposium on Low power electronics and design*. 2002, pp. 323–328, ACM Press.

[89] Alessandro Bogliolo, Roberto Corgnati, Enrico Macii, and Massimo Poncino, "Parameterized RTL Power Models for Soft Macros," *IEEE Transactions on VLSI*, vol. CAD-9, no. 6, pp. 880–887, Dec. 2001.

[90] T. Sato, Y. Ootaguro, M. Nagamatsu, and H. Tago, "Evaluation of architecture-level power estimation for cmos risc processors," in *Proc. Symp. Low Power Electronics*, Oct. 1995, pp. 44–45.

[91] P. Ong and R. Ynn, "Power-conscious software design – a framework for modeling software on hardware," in *Proc. of Symp. on Low Power Electronics*, 1994, pp. 36–37.

[92] David M. Brooks, Pradip Bose, Stanley E. Schuster, Hans Jacobson, Prabhakar N. Kudva, Alper Buyuktosunoğlu, John-David Wellman, Victor Zyuban, Manish Gupta, and Peter W. Cook, "Power-aware microarchitecture: Design and modeling challenges for next-generation microprocessors," *IEEE MICRO*, vol. 20, no. 6, pp. 26–44, Nov./Dec. 2000.

[93] W. Ye, N. Vijaykrishnan, M. Kandemir, and M. Irwin, "The design and use of simplepower: A cycleaccurate energy estimation tool," in *Proc. 37th Design Automation Conference*, Los Angeles, CA, June 2000.

[94] David Brooks, Vivek Tiwari, and Margaret Martonosi, "Wattch: a framework for architectural-level power analysis and optimizations," in *Proceedings ISCA 2000*, 2000, pp. 83–94.

[95] Russ Joseph and Margaret Martonosi, "Run-time power estimation in high-performance microprocessors," in *Proceedings of the International Symposium on Low Power Electronics and Design*, August 2001.

[96] David Brooks, John-David Wellman, Pradip Bose, and Margaret Martonosi, "Power-performance modeling and tradeoff analysis for a high-end microprocessor," in *Workshop on Power-Aware Computer Systems (PACS2000, held in conjuction with ASPLOS-IX)*, Nov. 2000.

[97] Gurhan Kucuk, Dmitry Ponomarev, and Kanad Ghose, "Accupower: an accurate power estimation tool for superscalar microprocessors," in *Design, Automation and Test in Europe Conference and Exhibition, 2002. Proceedings*, 2002, pp. 124–129.

[98] V. Tiwari, S. Malik, and A. Wolfe, "Power analysis of embedded software: a first step towards software power minimization," *IEEE Transactions on Very Large Scale Integration (VLSI) Systems*, vol. 2, no. 4, pp. 437–445, 1994.

[99] T. Lee, V. Tiwari, and S. Malik, "Power analysis and minimization techniques for embedded dsp software," *IEEE Transactions on VLSI Systems*, vol. 5, no. 1, pp. 123–135, 1997.

[100] V. Tiwari, S. Malik, A. Wolfe, and M. Lee, "Instruction level power analysis and optimization of software," *J. VLSI Signal Processing*, pp. 1–18, 1996.

[101] J. Russell and M. Jacome, "Software power estimation and optimization for high performance, 32-bit embedded processors," in *International Conference on Computer Design: VLSI in Computers and Processors*, 1998, pp. 328–333.

[102] D. Trifone D. Sarta and G. Ascia, "A data dependent approach to instruction level power estimation," in *Proc. IEEE Alessandro Volta Memorial Workshop on Low Power Design*, Como, Italy, March 1999, pp. 182–190.

[103] B. Klass, D. Thomas, H. Schmit, and D. Nagle, "Modeling inter-instruction energy effects in a digital signal processor," in *Power-Driven Microarchitecture Workshop*, June 1998.

[104] T. K. Tan, A. K. Raghunathan, G. Lakishminarayana, and N. K. Jha, "High-level software energy macro-modeling," in *Proceedings of the 38th conference on Design automation*. 2001, pp. 605–610, ACM Press.

[105] David Lidsky and Jan M. Rabaey, "Early power exploration - a world wide web application," in *Design Automation Conference*, 1996, pp. 27–32.

[106] D. Liu and C. Svensson, "Power Consumption Estimation in CMOS VLSI Chips," *IEEE Journal of Solid-State Circuits*, vol. 29, no. 6, pp. 663–669, June 1994.

[107] Raul San Martin and John P. Knight, "Power-profiler: Optimizing ASICs power consumption at the behavioral level," in *Design Automation Conference*, 1995, pp. 42–47.

[108] L. Benini, R. Hodgson, and P. Siegel, "System-level estimation and optimization," in *Proceedings of International Symposium of Low Power Electronics and Design Conference*, Aug 1998, pp. 173–178.

[109] A. Chandrakasan and R. Brodersen, "Minimizing power consumption in digital cmos circuits," *Proceedings of the IEEE*, vol. 83, no. 4, pp. 498–523, Apr. 1995.

[110] P. Landman, "High-level power estimation," in *Proc. ISLPED*, Monterey, CA, Aug. 1996, pp. 29–35.

[111] M. Sami, D. Sciuto, C. Silvano, and V. Zaccaria, "Instruction level power estimation for embedded VLIW cores," in *Proceedings of the 8th International Workshop on Hardware/Software Codesign (CODES-00)*, NY, May 3–5 2000, pp. 34–38, ACM.

[112] Mariagiovanna Sami, Donatella Sciuto, Cristina Silvano, and Vittorio Zaccaria, "Power exploration for embedded VLIW architectures," in *Proceedings of the IEEE/ACM International Conference on Computer Aided Design (ICCAD-2000)*, Nov. 5–9 2000, pp. 498–503.

[113] M. Sami, D. Sciuto, C. Silvano, and V. Zaccaria, "An Instruction-Level Energy Model for Embedded VLIW Architectures," *IEEE Transactions on Computer-Aided Design of Integrated Circuits and Systems*, vol. 21, no. r92, pp. 998–1010, September 2002.

[114] Andrea Bona, Mariagiovanna Sami, Donatella Sciuto, Cristina Silvano, Vittorio Zaccaria, and Roberto Zafalon, "Energy Estimation and Optimization of Embedded VLIW Processors Based on Instruction Clustering," in *Proceedings of the 39th Design Automation Conference DAC'02*, June 2002, pp. 886–891.

[115] E. Macii, "Sequential synthesis and optimization for low power," in *Low Power Design in Deep Submicron Electronics, NATO ASI Series, Series E: Applied Sciences*, 1997, pp. 321–353.

[116] P. Faraboschi, G. Brown, J. Fisher, G. Desoli, and F. Homewood, "Lx: a technology platform for customizable vliw embedded processing," in *Proceedings of the International Symposium on Computer Architecture*, June 2000, pp. 203–213.

[117] Chingren Lee, Jenq Kuen Lee, and TingTing Hwang, "Compiler optimization on instruction scheduling for low power," in *Proceedings of The 13th International Symposium on System Synthesis*. Sept. 20–22 2000, pp. 55–60, IEEE Computer Society Press.

[118] A. K. Jain, M. N. Murty, and P. J. Flynn, "Data clustering: a review.," *ACM Comp. Surveys*, vol. 31, no. 3, pp. 264–323, Sept. 1999.

[119] John H. Gennari, "A survey of clustering methods," Technical Report ICS-TR-89-38, University of California, Irvine, Department of Information and Computer Science, Oct. 1989.

[120] P. Geoffrey Lowney, Stefan M. Freudenberger, Thomas J. Karzes, W. D. Lichtenstein, Robert P. Nix, John S. O'Donnell, and John C. Ruttenberg, "The Multiflow Trace Scheduling compiler," *The Journal of Supercomputing*, vol. 7, no. 1-2, pp. 51–142, 1993.

[121] C. Lee, M. Potkonjak, and W. H. Mangione-Smith, "Mediabench: A tool for evaluating multimedia and communication systems," in *Proceedings of Micro 30*, 1997.

[122] S. Wilton and N. Jouppi, "CACTI:An Enhanced Cache Access and Cycle Time Model," *IEEE Journal of Solid-State Circuits*, vol. 31, no. 5, pp. 677–688, 1996.

[123] N. Vijaykrishnan, M. Kandemir, M.J. Irwin, H.S. Kim, and W. Ye, "Energy-driven integrated hardware-software optimizations using simplepower," in *ISCA 2000: 2000 International Symposium on Computer Architecture*, Vancouver BC, Canada, June 2000.

[124] Victor Zyuban and P. Kogge, "The energy complexity of register files," in *Proceedings of the International Symposium on Low Power Electronics and Design (ISLPED-98)*, New York, Aug. 10–12 1998, pp. 305–310, ACM Press.

[125] M. B. Kamble and K. Ghose, "Analytical energy dissipation models for low power caches," in *ISLPED-97: ACM/IEEE Int. Symposium on Low Power Electronics and Design*, 1997.

[126] L. Benini, D. Bruni, M. Chinosi, C. Silvano, V. Zaccaria, and R. Zafalon, "A power modeling and estimation framework for vliw-based embedded systems," in *Proceedings of International Workshop-Power And Timing Modeling, Optimization and Simulation, PATMOS'01*, 26–28 2001.

[127] L. Benini, D. Bruni, M. Chinosi, C. Silvano, V. Zaccaria, and R. Zafalon, "A power modeling and estimation framework for vliw-based embedded systems," in *Proceedings of International Workshop-Power And Timing Modeling, Optimization and Simulation, PATMOS'01*, 26–28 2001.

[128] T. D. Burd and R. W. Brodersen, "Energy efficient CMOS microprocessor design," in *In Proc. 28th Hawaii Int. Conf. on System Sciences*, Jan. 1995, pp. 288–297.

[129] M. Pedram, "Power Minimization in IC Design: Principles and Applications," *ACM Transactions on Design Automation of Electronic Systems (Todaes)*, vol. 1, no. 1, pp. 3–56, Jan. 1996.

[130] R. Brayton, R. Rudell, and A. Sangiovanni-Vincentelli, "Mis: A multiple-level logic optimization," *IEEE Transaction on CAD of Integrated Circuits and Systems*, pp. 1062–1081, Nov. 1987.

[131] K. Roy and S. Prasad, "SYCLOP: Synthesis of CMOS logic for low power applications," in *Proc. of ICCAD-92*, 1992, pp. 464–467.

[132] Sasan Iman and Massoud Pedram, "Logic extraction and factorization for low power," in *Proceedings of the 32nd ACM/IEEE conference on Design automation conference*. 1995, pp. 248–253, ACM Press.

[133] K. Keutzer, "Dagon: Technology binding and local optimization by dag matching," in *Proc. 24th DAC*, 1987, pp. 341–347.

[134] C. Tsui, M. Pedram, and A. Despain, "Technology decomposition and mapping targeting low power dissipation," in *Proc. Design Automation Conf.*, 1993, pp. 68–73.

[135] V. Tiwari, "Technology mapping for low power in logic synthesis," *The VLSI Journal*, vol. 20, no. 3, 1996.

[136] B. Lin and H. De Man, "Low-power driven technology mapping under timing constraints," in *Proceedings of the ICCD'93*, 1993, pp. 421–427.

[137] L. Benini and G. De Micheli, "Automatic synthesis of gated-clock sequential circuits," *IEEE Trans. Computer-Aided Design of Integrated Circuits and Systems*, vol. 15, no. 6, pp. 630–643, June 1996.

[138] L. Benini, P. Siegel, and G. De Micheli, "Saving Power by Synthesizing Gated Clocks for Sequential Circuits," *IEEE J. of Design and Test of Computers*, 1994.

[139] L. Benini and G. De Micheli, "Transformation and synthesis of FSM's for low power gated clock implementation," *IEEE Trans. Computer Aided Design of Integrated Circuits*, vol. 15, no. 6, pp. 630–643, 1996.

[140] L. Benini, G. De Micheli, A. Macii, E. Macii, M. Poncino, and R. Scarsi, "Glitch power minimization by selective gate freezing," *IEEE Transactions on Very Large Scale Integration (VLSI) Systems*, vol. 8, no. 3, pp. 287–298, June 2000.

[141] L. Benini, M. Favalli, and G. De Micheli, "Design for testability of gated-clock fsm's," in *In Proc. EDTC-96: IEEE Eur. Design and Test Conf.*, Paris, France, Mar. 1996, pp. 589–596.

[142] M. Alidina, J. Monteiro, S. Devadas, and A. Ghosh, "Precomputation-based sequential logic optimization for low power," *IEEE Transactions on VLSI Systems*, vol. 2, no. 4, pp. 426–436, 1994.

[143] L. Benini and G. De Micheli, "State assignment for Low Power Dissipation," *IEEE Journal of Solid State Circuits*, vol. 30, no. 3, pp. 258–268, 1994.

[144] C.-Y. Tsui, M. Pedram, C.-A. Chen, and A. M. Despain, "Low power state assignment targeting two- and multi-level logic implementations," in *International Conference on Computer-Aided Design*, Los Alamitos, Ca., USA, Nov. 1994, pp. 82–89, IEEE Computer Society Press.

[145] E. Olson and S. M. Kang, "Low-power state assignment for finite state machines," in *Proc. ACM/IEEE IWLPD-94*, Napa Valley, CA, Apr. 1994, pp. 63–68.

[146] P. Surti, L. F. Chao, and A. Tyagi, "Low power fsm design using huffman-style encoding," in *Proc. EDTC-97*, Paris, France, Mar. 1994, pp. 521–525.

[147] Chi-Ying Tsui, Massoud Pedram, and Alvin M. Despain, "Exact and approximate methods for calculating signal and transition probabilities in FSMs," in *Proceedings of the 31st Conference on Design Automation*, Michael Lorenzetti, Ed., New York, NY, USA, June 1994, pp. 18–23, ACM Press.

[148] C. E. Leiserson and J. B. Saxe, "Retiming Synchronous Circuit," Technical Memo MIT/LCS/TM-372, Massachusetts Institute of Technology, Laboratory for Computer Science, Oct. 1988.

[149] J. Monteiro, S. Devadas, and A. Ghosh, "Retiming sequential circuits for low power," in *Proceedings of the IEEE/ACM International Conference on*

Computer-Aided Design, Michael Lightner, Ed., Santa Clara, CA, Nov. 1993, pp. 398–402, IEEE Computer Society Press.

[150] J. K. Kin, M. Gupta, and W. H. Mangione-Smith, "Filtering Memory References to Increase Energy Efficiency," *IEEE Trans. on Computers*, vol. 49, no. 1, Jan. 2000.

[151] R. Gonzalez and M. Horowitz, "Energy dissipation in general purpose microprocessors," *IEEE Journal of Solid-State Circuits*, vol. 31, no. 9, pp. 1277–1284, Sept. 1996.

[152] R. Canal, A. Gonzalez, and J. Smith, "Very low power pipelines using significance compression," in *Proc. Micro-33*, Dec. 2000, pp. 181–190.

[153] J. Brennan, A. Dean, S. Kenyon, and S. Ventrone, "Low power methodology and design techniques for processor design," in *Proc. of the 1998 Int'l. Symp. on Low Power Electronics and Design*, Monterey, California, Aug. 1998, pp. 268–273.

[154] M. Hiraki, "Stage-skip pipeline: A low power processor architecture using a decoded instruction buffer," in *Proc. 1996 International Symposium on Low Power Electronics and Design*, Aug. 1996.

[155] Luca Benini, Giovanni de Micheli, Alberto Macii, Enrico Macii, and Massimo Poncino, "Automatic selection of instruction op-codes of low-power core processors," *Computers and Digital Techniques, IEE Proceedings*, vol. 146, no. 4, pp. 173–178, July. 1999.

[156] M. Stan and W. Burleson, "Bus-Invert Coding for Low-Power I/O," *IEEE Trans. on VLSI Systems*, pp. 49–58, Mar. 1995.

[157] C. Su, C. Tsui, and A. Despain, "Saving power in the control path of embedded processors," *IEEE Design and Test of computers*, vol. 11, pp. 24–30, 1994.

[158] L. Benini, G. Micheli, E. Macii, D. Sciuto, and C. Silvano, "Asymptotic zero-transition activity encoding for address buses in low-power microprocessor-based systems," in *Proc. Great Lake Symposium on VLSI*, Urbana, IL, 1997.

[159] L. Benini, G. De Micheli, E. Macii, D. Sciuto, and C. Silvano, "Address bus encoding techniques for system-level power optimization," in *Proc. Design, Automation and Test in Europe*, Feb. 1998, pp. 861–866.

[160] W. Fornaciari, M. Polentarutti, D. Sciuto, and C. Silvano, "Power optimization of system-level address buses based on software profiling," in *CODES-2000: 8th Int. Workshop on Hardware/Software Co-Design*, San Diego, CA, May 2000.

[161] E. Musoll, T. Lang, and J. Cortadella, "Exploiting the locality of memory references to reduce the address bus energy," in *In Int. Symp. on Low Power Design and Electronics*, Aug. 1997, pp. 202–207.

[162] L. Benini, G. De Micheli, E. Macii, M. Poncino, and S. Quez, "System-level power optimization of special purpose applications: the beach solution," in

Proceedings of the International Symposium on Low Power Electronics and Design, 1997, pp. 24–29.

[163] Luca Benini, Giovanni de Micheli, Enrico Macii, Massimo Poncino, and Stefano Quer, "Power Optimization of Core-Based Systems by Address Bus Encoding," *IEEE Transactions on VLSI Systems*, vol. 6, no. 4, pp. 554–562, Dec. 1998.

[164] Luca Benini, Alberto Macii, Enrico Macii, Massimo Poncino, and Riccardo Scarsi, "Stream synthesis for efficient power simulation based on spectral transforms," *IEEE Transactions on CAD of Integrated Circuits and Systems*, vol. 19, no. 9, pp. 969–980, Sept. 2000.

[165] M. Johnson, *Superscalar Microprocessor Design*, Prentice-Hall, Englewood Cliffs, New Jersey, 1991.

[166] C-L. Su, C-Y. Tsui, and A. M. Despain, "Low power architecture design and compilation techniques for high-performance processors.," Tech. Rep. ACAL-TR-94-01, Univ. of Southern California, Feb. 1994, Advanced Computer Architecture Laboratory.

[167] M. Toburen, T. Conte, and M. Reilly, "Instruction scheduling for low power disspiation in high performance microprocessors," in *Proceedings of the Power Driven Microarchitecture Workshop*, June 1998.

[168] J. Chang and M. Pedram, "Register allocation and binding for low power," in *Proceedings of the Annual ACM IEEE Design Automation Conference*, June 1995, pp. 29–35.

[169] H. Mehta, R. Owens, M. Irwin, R. Chen, and D. Ghosh, "Techniques for low energy software," in *Proceedings of the International Symposium on Low Power Electronics and Design*, August 1997, pp. 72–75.

[170] Mahmut T. Kandemir, N. Vijaykrishnan, Mary Jane Irwin, and W. Ye, "Influence of compiler optimizations on system power.," in *Proceedings of the 37th Conference on Design Automation (DAC-00)*, NY, June 5–9 2000, pp. 304–307, ACM/IEEE.

[171] A. Parikh, M. Kandemir, N. Vijaykrishnan, and M.J. Irwin, "Vliw scheduling for energy and performance," in *IEEE Computer Society Workshop on VLSI*, Orlando, Florida, Apr. 2001.

[172] N. Zervas, K. Masselos, and C. Goutis, "Code transformations for embedded multimedia applications: Impact of power and performance," in *Proceedings of the Power-Driven Microarchitecture Workshop In Conjunction With ISCA98*, June 1998.

[173] A. Parikh, M. Kandemir, N. Vijaykrishnan, and M.J. Irwin, "Instruction scheduling based on energy and performance constraints," in *Proceedings of the IEEE Computer Society Workshop on VLSI*. Apr. 27–28 2000, pp. 37–42, IEEE Computer Society Press.

[174] E. Brockmeyer, L. Nachtergaele, F. Catthoor, J. Bormans, and H. De Man, "Low power memory storage and transfer organization for the MPEG-4 full pel

motion estimation on a multi media processor," *IEEE Trans. on Multi-Media*, vol. 1, no. 2, pp. 202–216, 1999.

[175] F. Catthoor, S. Wuytack, E. De Greef, F. Balasa, and P. Slock, *System exploration for custom low power data storage and transfer*, Marcel Dekker, Inc., New York, 1998.

[176] F. Catthoor, S. Wuytackand E. De Greef, F. Franssen, L. Nachtergaele, and H. De Man, "System-level transformations for low power data transfer and storage," in *Low power CMOS design*, R. Brodersen A. Chandrakasan, Ed., pp. 609–618. IEEE Press, 1998.

[177] K. Danckaert, K. Masselos, F. Catthoor, H. De Man, and C. Goutis, "Strategy for power efficient design of parallel systems," *IEEE Trans. on VLSI Systems*, vol. 7, no. 2, pp. 258–265, 1999.

[178] P.R. Panda and N.D. Dutt, "Low-power memory mapping through reducing address bus activity," *IEEE Trans. on VLSI Systems*, vol. 7, no. 3, pp. 309–320, 1999.

[179] S. Udayanarayanan and C. Chakrabarti, "Energy-efficient code generation for DSP56000 family," in *Proceedings of the 2000 International Symposium on Low Power Electronics and Design (ISLPED-00)*, N. Y., July 26–27 2000, pp. 247–249, ACM Press.

[180] Y. Li and J. Henkel, "A framework for estimating and minimizing energy dissipation of embedded hw/sw systems," in *DAC-35: ACM/IEEE Design Automation Conference*, June 1998.

[181] T. M. Conte, K. N. Menezes, S. W. Sathaye, and M. C. Toburen, "System-level power consumption modeling and tradeoff analysis techniques for superscalar processor design," *IEEE Trans. on Very Large Scale Integration (VLSI) Systems*, vol. 8, no. 2, pp. 129–137, Apr. 2000.

[182] N. Bellas, I. N. Hajj, D. Polychronopoulos, and G. Stamoulis, "Architectural and compiler techniques for energy reduction in high-performance microprocessors," *IEEE Transactions on Very Large Scale of Integration (VLSI) Systems*, vol. 8, no. 3, June 2000.

[183] R. I. Bahar, G. Albera, and S. Manne, "Power and performance tradeoffs using various caching strategies," in *ISLPED-98: ACM/IEEE Int. Symposium on Low Power Electronics and Design*, Monterey, CA, 1998.

[184] Tony D. Givargis, Frank Vahid, and Jrg Henkel, "Evaluating power consumption of parameterized cache and bus architectures in system-on-a-chip designs," *IEEE Transactions on Very Large Scale of Integration (VLSI) Systems*, vol. 9, no. 4, August 2001.

[185] L. Benini, A. Macii, E. Macii, , and M. Poncino, "Increasing energy efficiency of embedded systems by application-specific memory hierarchy generation," *Design and Test of Computers*, vol. 17, no. 2, pp. 74–85, April–June 2000.

[186] C. L. Su and A. M. Despain, "Cache design trade-offs for power and performance optimization: A case study," in *ISLPED-95: ACM/IEEE Int. Symposium on Low Power Electronics and Design*, 1995.

[187] S. E. Wilton and N. Jouppi, "An enhanced access and cycle time model for on-chip caches," Tech. Rep. 93/5, Digital Equipment Corporation Western Research Lab., 1994.

[188] P. Hicks, M. Walnock, and R. M. Owens, "Analysis of power consumption in memory hierarchies," in *ISLPED-97: ACM/IEEE Int. Symposium on Low Power Electronics and Design*, Monterey, CA, 1997, pp. 239–242.

[189] W.-T. Shiue and C. Chakrabarti, "Power estimation of system-level buses for microprocessor-based architectures: A case study," in *Proc. DAC99: Design Automation Conference*, New Orleans, LU, 1999.

[190] Luca Benini, Alberto Macii, Enrico Macii, and Massimo Poncino, "Increasing energy efficiency of embedded systems by application-specific memory hierarchy generation," *IEEE Design & Test of Computers*, vol. 17, no. 2, pp. 74–85, Apr. 2000.

[191] Luca Benini and Giovanni de Micheli, "System-level power optimization: techniques and tools," *ACM Transactions on Design Automation of Electronic Systems.*, vol. 5, no. 2, pp. 115–192, Jan. 2000.

[192] L. Benini, A. Bogliolo, and G. Micheli, "A Survey of Design Techniques for System-Level Dynamic Power Management," *IEEE Transactions on VLSI Systems*, Mar. 2000.

[193] Anna R. Karlin, Mark S. Manasse, Lyle A. McGeoch, and Susan S. Owicki, "Competitive randomized algorithms for nonuniform problems," *Algorithmica*, vol. 11, no. 6, pp. 542–571, June 1994.

[194] Richard Golding, Peter Bosch, Carl Staelin, Tim Sullivan, and John Wilkes, "Idleness is not sloth," in *Proceedings of the 1995 USENIX Technical Conference: January 16–20, 1995, New Orleans, Louisiana, USA*, USENIX Association, Ed., Berkeley, CA, USA, Jan. 1995, pp. 201–212, USENIX.

[195] M. Srivastava, A. Chandrakasan, and R. Broderson, "Predictive Shutdown and Other Architectural Techniques for Energy Efficient Programmable Computation," *IEEE Trans. on VLSI Systems*, vol. 4, no. 1, pp. 42–54, March 1996.

[196] C.-H. Hwang and A. C.-H. Wu, "A predictive system shutdown method for energy saving of event-driven computation," in *IEEE/ACM International Conference on Computer Aided Design; Digest of Technical Papers (ICCAD '97)*, Washington - Brussels - Tokyo, Nov. 1997, pp. 28–32, IEEE Computer Society Press.

[197] Fred Douglis, P. Krishnan, and Brian Bershad, "Adaptive disk spin-down policies for mobile computers," in *Proceedings of the second USENIX Symposium on Mobile and Location-Independent Computing: April 10–11, 1995, Ann*

Arbor, Michigan, USA, USENIX Association, Ed., Berkeley, CA, USA, Apr. 1995, pp. 121–137, USENIX.

[198] David P. Helmbold, Darrell D. E. Long, and Bruce Sherrod, "A dynamic disk spin-down technique for mobile computing," in *Proc. Int. Conf. on Mobile Computing and Networking*, 1996, pp. 130–142.

[199] Y.-H. Lu, G. De Micheli, and L. Benini, "Operating-system directed power reduction," in *Proceedings of the 2000 International Symposium on Low Power Electronics and Design (ISLPED-00)*, N. Y., July 26–27 2000, pp. 37–42, ACM Press.

[200] G. A. Paleologo, Luca Benini, Alessandro Bogliolo, and Giovanni De Micheli, "Policy optimization for dynamic power management," in *Design Automation Conference*, 1998, pp. 182–187.

[201] Vittorio Zaccaria, "Workload Characterization for Dynamic Power Management," Tech. Rep. 2000.42, Dipartimento di Elettronica e Informazione. Politecnico di Milano, november 2000.

[202] A. Bogliolo Eui-Young, L. Benini and G. De Micheli, "Dynamic power management for nonstationary service requests," in *Design, Automation and Test in Europe Conference and Exhibition 1999. Proceedings*, N. Y., July 26–27 1999, pp. 77 – 81, ACM Press.

[203] Henk Corporaal, *Microprocessor Architectures – from VLIW to TTA*, John Wiley & Sons, 1998.

[204] Arthur Abnous and Nader Bagherzadeh, "Pipelining and bypassing in a VLIW processor," *IEEE Transactions on Parallel and Distributed Systems*, vol. 5, no. 6, pp. 658–664, June 1994.

[205] R. Yung and N. C. Wilhelm, "Caching processor general registers," in *Proc. Intl. Conf. Computer Design*, Oct. 1995, pp. 307–312.

[206] Luis A. Lozano C. and Guang R. Gao, "Exploiting short-lived variables in superscalar processors," in *Proceedings of the 28th Annual International Symposium on Microarchitecture*, Ann Arbor, Michigan, Nov. 29–Dec. 1, 1995, IEEE Computer Society TC-MICRO and ACM SIGMICRO, pp. 292–302.

[207] M. M. Martin, A. Roth, and C. N. Fischer, "Exploiting dead value information," in *Proceedings of the 30th Annual IEEE/ACM International Symposium on Microarchitecture (MICRO-97)*, Los Alamitos, Dec. 1–3 1997, pp. 125–137, IEEE Computer Society.

[208] Alfred V. Aho, Ravi Sethi, and Jeffrey D. Ullman, *Compilers: Principles, Techniques, and Tools*, Addison-Wesley, Reading, MA, USA, 1986.

[209] N. H. Chen, C. H. Smith, and S. C. Fralick, "A fast computational algorithm for discrete cosine transform," *IEEE Transactions on Communications*, vol. COM-25, no. 9, Sept. 1977.

[210] William H. Press, Saul A. Teukolsky, William T. Vetterling, and Brian P. Flannery, *Numerical Recipes in C: The Art of Scientific Computing*, Cambridge University Press, Cambridge, 2 edition, 1992.

[211] W. Fornaciari, D. Sciuto, C. Silvano, and V. Zaccaria, "Fast system-level exploration of memory architecures driven by energy-delay metrics," in *Proceedings of ISCAS-2001: International Symposium on Circuits and Systems*, May 6–9 2001, vol. IV, pp. 502–506.

[212] W. Fornaciari, D. Sciuto, C. Silvano, and V. Zaccaria, "A design framework to efficiently explore energy-delay tradeoffs," in *Proceedings of CODES-2001: Ninth International Symposium on Hardware/Software Codesign*, Apr. 25–27 2001, pp. 260–265.

[213] NEC, "16m-bit synchronous dram, doc. no. m12939ej3v0ds00," Data Sheet, April 1998.

MIX
Papier aus verantwortungsvollen Quellen
Paper from responsible sources
FSC® C105338

If you have any concerns about our products,
you can contact us on
ProductSafety@springernature.com

In case Publisher is established outside the EU,
the EU authorized representative is:
Springer Nature Customer Service Center GmbH
Europaplatz 3, 69115 Heidelberg, Germany

Printed by Libri Plureos GmbH
in Hamburg, Germany